Colliery Engineer Co.

The Elements of Railroad Engineering

Volume V

Colliery Engineer Co.

The Elements of Railroad Engineering
Volume V

ISBN/EAN: 9783744729116

Printed in Europe, USA, Canada, Australia, Japan

Cover: Foto ©berggeist007 / pixelio.de

More available books at **www.hansebooks.com**

THE ELEMENTS

OF

RAILROAD ENGINEERING

PREPARED FOR STUDENTS OF

THE INTERNATIONAL CORRESPONDENCE SCHOOLS

SCRANTON, PA.

Volume V

ANSWERS TO QUESTIONS

First Edition

SCRANTON

THE COLLIERY ENGINEER CO.

1897

BURR PRINTING HOUSE,
FRANKFORT AND JACOB STREETS,
NEW YORK.

A KEY

TO ALL THE

QUESTIONS AND EXAMPLES

INCLUDED IN

VOLS. I AND II,

EXCEPT THE

EXAMPLES FOR PRACTICE.

It will be noticed that the Key is divided into sections which correspond to the sections containing the questions and examples at the end of Vols. I and II. The answers and solutions are so numbered as to be similar to the numbers before the questions to which they refer.

CONTENTS.

ARITHMETIC.

(QUESTIONS 1-75.)

(1) See Art. **1.**

(2) See Art. **3.**

(3) See Arts. **5** and **6.**

(4) See Arts. **10** and **11.**

(5) 980 = Nine hundred eighty.

605 = Six hundred five.

28,284 = Twenty-eight thousand, two hundred eighty-four,

9,006,042 = Nine million, six thousand and forty-two.

850,317,002 = Eight hundred fifty million, three hundred seventeen thousand and two.

700,004 = Seven hundred thousand and four.

(6) Seven thousand six hundred = 7,600.

Eighty-one thousand four hundred two = 81,402.

Five million, four thousand and seven = 5,004,007.

One hundred and eight million, ten thousand and one = 108,010,001.

Eighteen million and six = 18,000,006.

Thirty thousand and ten = 30,010.

(7) In adding whole numbers, place the numbers to be added directly under each other so that the extreme right-hand figures will stand in the same column, regardless of the position of those at the left. Add the first column of figures at the extreme right, which equals 19 units, or 1 ten and 9 units. We place 9 units under the units column, and reserve 1 ten for the column

$$
\begin{array}{r}
3290 \\
504 \\
865403 \\
2074 \\
81 \\
7 \\
\hline
871359 \quad \text{Ans.}
\end{array}
$$

of tens. $1 + 8 + 7 + 9 = 25$ tens, or 2 hundreds and 5 tens. Place 5 tens under the tens column, and reserve 2 hundreds for the hundreds column. $2 + 4 + 5 + 2 = 13$ hundreds, or 1 thousand and 3 hundreds. Place 3 hundreds under the hundreds column, and reserve the 1 thousand for the thousands column. $1 + 2 + 5 + 3 = 11$ thousands, or 1 ten-thousand and 1 thousand. Place the 1 thousand in the column of thousands, and reserve the 1 ten-thousand for the column of ten-thousands. $1 + 6 = 7$ ten-thousands. Place this seven ten-thousands in the ten-thousands column. There is but one figure 8 in the hundreds of thousands place in the numbers to be added, so it is placed in the hundreds of thousands column of the sum.

A simpler (though less scientific) explanation of the same problem is the following: $7 + 1 + 4 + 3 + 4 + 0 = 19$; write the nine and reserve the 1. $1 + 8 + 7 + 0 + 0 + 9 = 25$; write the 5 and reserve the 2. $2 + 0 + 4 + 5 + 2 = 13$; write the 3 and reserve the 1. $1 + 2 + 5 + 3 = 11$; write the 1 and reserve 1. $1 + 6 = 7$; write the 7. Bring down the 8 to its place in the sum.

(8)
$$
\begin{array}{r}
709 \\
8304725 \\
391 \\
100302 \\
300 \\
909 \\
\hline
8407336 \quad \text{Ans.}
\end{array}
$$

(9) (*a*) In subtracting whole numbers, place the subtrahend or smaller number under the minuend or larger number, so that the right-hand figures stand directly under each other. Begin *at the right* to subtract. We can not subtract 8 units from 2 units, so we take 1 ten from the 6 tens and add it to the 2 units. As 1 *ten* = 10 *units*, we have 10 units + 2 units = 12 units. Then, 8 units from 12 units leaves 4 units. We took 1 ten from 6 tens, so

only 5 tens remain. 3 tens from 5 tens
leaves 2 tens. In the hundreds column we
have 3 hundreds from 9 hundreds leaves
6 hundreds. We can not subtract 3 thou-
sands from 0 thousands, so we take 1 ten-thousand from
5 ten-thousands and add it to the 0 thousands. 1 *ten-
thousand* = 10 *thousands*, and 10 thousands + 0 thousands
= 10 thousands. Subtracting, we have 3 thousands from
10 thousands leaves 7 thousands. We took 1 ten-thousand
from 5 ten-thousands and have 4 ten-thousands remaining.
Since there are no ten-thousands in the subtrahend, the
4 in the ten-thousands column in the minuend is brought
down into the same column in the remainder, because 0 from
4 leaves 4.

$$\begin{array}{r} 5\,0\,9\,6\,2 \\ 3\,3\,3\,8 \\ \hline 4\,7\,6\,2\,4 \quad \text{Ans.} \end{array}$$

(*b*)
$$\begin{array}{r} 1\,5\,3\,3\,9 \\ 1\,0\,0\,0\,1 \\ \hline 5\,3\,3\,8 \quad \text{Ans.} \end{array}$$

(**10**) (*a*)
$$\begin{array}{r} 7\,0\,9\,6\,8 \\ 3\,2\,9\,7\,5 \\ \hline 3\,7\,9\,9\,3 \quad \text{Ans.} \end{array}$$
(*b*)
$$\begin{array}{r} 1\,0\,0\,0\,0\,0 \\ 9\,8\,7\,3\,5 \\ \hline 1\,2\,6\,5 \quad \text{Ans.} \end{array}$$

(**11**) We have given the minuend or greater number
(1,004) and the difference or remainder (49). Placing these
in the usual form of subtraction we have $\dfrac{1\,0\,0\,4}{4\,9}$ in which
the dash (———) represents the number sought. This number
is evidently *less* than 1,004 by the difference 49, hence,
1,004 − 49 = 955, the smaller number. For the sum of the
two numbers we then have
$$\begin{array}{l} 1\,0\,0\,4 \ \textit{larger} \\ \underline{9\,5\,5} \ \textit{smaller} \\ 1\,9\,5\,9 \ \textit{sum.} \quad \text{Ans.} \end{array}$$
Or, this problem may be solved as follows: If the greater
of two numbers is 1,004, and the difference between them is
49, then it is evident that the smaller number must be
equal to the difference between the greater number (1,004)

and the difference (49); or, 1,004 − 49 = 955, the smaller number. Since the greater number equals 1,004 and the smaller number equals 955, their sum equals 1,004 + 955 = 1,959 sum.   Ans.

(12)   The numbers connected by the plus (+) sign must first be added.   Performing these operations we have

$$
\begin{array}{r}
5962 \\
8471 \\
\underline{9023} \\
23456 \; sum.
\end{array}
\qquad
\begin{array}{r}
3874 \\
\underline{2039} \\
5913 \; sum.
\end{array}
$$

Subtracting the smaller number (5,913) from the greater (23,456) we have

$$
\begin{array}{r}
23456 \\
\underline{5913} \\
17543 \; difference.
\end{array}
\quad \text{Ans.}
$$

(13) 44675 = amount willed to his son.
         $\underline{26380}$ = amount willed to his daughter.
         71055 = amount willed to his two children.
       125000 = amount willed to his wife and two children.
         $\underline{71055}$ = amount willed to his two children.
       53945 = amount willed to his wife.   Ans.

(14)   In the multiplication of whole numbers, place the multiplier under the multiplicand, and multiply each term of the multiplicand by each term of the multiplier, writing the right-hand figure of each product obtained under the term of the multiplier which produces it.

(a)   7 × 7 units = 49 units, or 4 tens and 9
526387   units.   We write the 9 units and reserve
$\underline{7}$ the 4 tens.   7 times 8 tens = 56 tens;
3684709   Ans.   56 tens + 4 tens reserved = 60 tens or
                    6 hundreds and 0 tens.   Write the 0
tens and reserve the 6 hundreds.   7 × 3 hundreds = 21 hundreds; 21 + 6 hundreds reserved = 27 hundreds, or 2 thousands and 7 hundreds.   Write the 7 hundreds and reserve

the 2 thousands. 7×6 thousands $= 42$ thousands; 42 $+ 2$ thousands reserved $= 44$ thousands or 4 ten-thousands and 4 thousands.   Write the 4 thousands and reserve the 4 ten-thousands. 7×2 ten-thousands $= 14$ ten-thousands; $14 + 4$ ten-thousands reserved $= 18$ ten-thousands, or 1 hundred-thousand and 8 ten-thousands.   Write the 8 ten-thousands and reserve the 1 hundred-thousand. 7×5 hundred-thousands $= 35$ hundred-thousands; $35 + 1$ hundred-thousand reserved $= 36$ hundred-thousands.   Since there are no more figures in the multiplicand to be multiplied, we write the 36 hundred-thousands in the product.   This completes the multiplication.

A simpler (though less scientific) explanation of the same problem is the following:

7 times $7 = 49$; write the 9 and reserve the 4.   7 times $8 = 56$; $56 + 4$ reserved $= 60$; write the 0 and reserve the 6. 7 times $3 = 21$; $21 + 6$ reserved $= 27$; write the 7 and reserve the 2. $7 \times 6 = 42$; $42 + 2$ reserved $= 44$; write the 4 and reserve 4. $7 \times 2 = 14$; $14 + 4$ reserved $= 18$; write the 8 and reserve the 1. $7 \times 5 = 35$; $35 + 1$ reserved $= 36$; write the 36.

In this case the multiplier is 17 *units*, or 1 *ten* and 7 *units*, so that the product is obtained by adding two partial products, namely, $7 \times$ 700,298 and $10 \times 700,298$.   The actual operation is performed as follows:

$$(b) \quad \begin{array}{r} 700298 \\ 17 \\ \hline 4902086 \\ 700298 \\ \hline 11905066 \end{array} \text{ Ans.}$$

7 times $8 = 56$; write the 6 and reserve the 5.   7 times $9 = 63$; $63 + 5$ reserved $= 68$; write the 8 and reserve the 6. 7 times $2 = 14$; $14 + 6$ reserved $= 20$; write the 0 and reserve the 2.   7 times $0 = 0$; $0 + 2$ reserved $= 2$; write the 2. 7 times $0 = 0$; $0 + 0$ reserved $= 0$; write the 0.   7 times $7 = 49$; $49 + 0$ reserved $= 49$; write the 49.

. To multiply by the 1 ten we say 1 times $700298 = 700298$, and write 700298 under the first partial product, as shown, with the right-hand figure 8 under the multiplier 1.   Add the two partial products; their sum equals the entire product.

(*c*)   2 1 7   Multiply any two of the numbers together
       1 0 3   and multiply their product by the third
       ─────   number.
       6 5 1
     2 1 7 0
     ───────
     2 2 3 5 1
           6 7
     ─────────
     1 5 6 4 5 7
   1 3 4 1 0 6
   ───────────
   1 4 9 7 5 1 7   Ans.

(15) If your watch ticks every second, then to find how many times it ticks in one week it is necessary to find the number of seconds in 1 week.

    6 0 seconds = 1 minute.
    6 0 minutes = 1 hour.
    ─────
    3 6 0 0 seconds = 1 hour.
       2 4 hours = 1 day.
    ─────
    1 4 4 0 0
    7 2 0 0
    ─────────
    8 6 4 0 0 seconds = 1 day.
         7 days = 1 week.
    ─────────
    6 0 4 8 0 0 seconds in 1 week or the number of times that
    Ans.   your watch ticks in 1 week.

(16) If a monthly publication contains 24 pages, a yearly
     2 4   volume will contain 12×24 or 288 pages, since
     1 2   there are 12 months in one year; and eight
     ───   yearly volumes will contain 8×288, or 2,304
     2 8 8   pages.
         8
     ─────
     2 3 0 4   Ans.

(17) If an engine and boiler are worth $3,246, and the building is worth 3 times as much, plus $1,200, then the building is worth

                $ 3 2 4 6
                        3
                ─────────
                9 7 3 8
        plus   1 2 0 0
                ─────────
                $ 1 0 9 3 8 = value of building.

If the tools are worth twice as much as the building, plus
$1,875, then the tools are worth

$$\$10938$$
$$2$$

$$\overline{21876}$$
$$plus \quad 1875$$

$$\overline{\$23751} = \text{value of tools.}$$

Value of building = $10938
Value of tools   = 23751

$$\overline{\$34689} = \text{value of the building}$$
$$\text{and tools. } (a) \text{ Ans.}$$

Value of engine and
boiler = $ 3246
Value of building
and tools = 34689

$$\overline{\$37935} = \text{value of the whole}$$
$$\text{plant. } (b) \text{ Ans.}$$

(18) (a) $(72 \times 48 \times 28 \times 5) \div (96 \times 15 \times 7 \times 6)$.
Placing the numerator over the denominator the problem
becomes

$$\frac{72 \times 48 \times 28 \times 5}{96 \times 15 \times 7 \times 6} = ?$$

The 5 in the *dividend* and 15 in the *divisor* are both *divisible* by 5, since 5 divided by 5 equals 1, and 15 divided by
5 equals 3. *Cross off* the 5 and write the 1 *over* it; also *cross off* the 15 and write the 3 *under* it. Thus,

$$\frac{72 \times 48 \times 28 \times \overset{1}{\cancel{5}}}{96 \times \underset{3}{\cancel{15}} \times 7 \times 6} =$$

The 5 and 15 are *not* to be considered any longer, and, in
fact, may be erased entirely and the 1 and 3 placed in their
stead, and treated as if the 5 and 15 *never* existed. Thus,

$$\frac{72 \times 48 \times 28 \times 1}{96 \times 3 \times 7 \times 6} =$$

72 in the *dividend* and 96 in the *divisor* are *divisible* by 12, since 72 divided by 12 equals 6, and 96 divided by 12 equals 8. *Cross off* the 72 and write the 6 *over* it; also, *cross off* the 96 and write the 8 *under* it. Thus,

$$\frac{\overset{6}{\cancel{72}} \times 48 \times 28 \times 1}{\underset{8}{\cancel{96}} \times 3 \times 7 \times 6} =$$

The 72 and 96 are *not* to be considered any longer, and, in fact, may be *erased* entirely and the 6 and 8 placed in their stead, and treated as if the 72 and 96 *never* existed. Thus,

$$\frac{6 \times 48 \times 28 \times 1}{8 \times 3 \times 7 \times 6} =$$

Again, 28 in the *dividend* and 7 in the *divisor* are *divisible* by 7, since 28 divided by 7 equals 4, and 7 divided by 7 equals 1. *Cross off* the 28 and write the 4 *over* it; also, cross off the 7 and write the 1 *under* it. Thus,

$$\frac{6 \times 48 \times \overset{4}{\cancel{28}} \times 1}{8 \times 3 \times \underset{1}{\cancel{7}} \times 6} =$$

The 28 and 7 are *not* to be considered any longer, and, in fact, may be *erased* entirely and the 4 and 1 placed in their stead, and treated as if the 28 and 7 *never* existed. Thus,

$$\frac{6 \times 48 \times 4 \times 1}{8 \times 3 \times 1 \times 6} =$$

Again, 48 in the *dividend* and 6 in the *divisor* are *divisible* by 6, since 48 divided by 6 equals 8, and 6 divided by 6 equals 1. *Cross off* the 48 and write the 8 *over* it; also, cross off the 6 and write the 1 *under* it. Thus,

$$\frac{6 \times \overset{8}{\cancel{48}} \times 4 \times 1}{8 \times 3 \times 1 \times \underset{1}{\cancel{6}}} =$$

The 48 and 6 are *not* to be considered any longer, and, in fact, may be *erased* entirely and the 8 and 1 placed in their stead, and treated as if the 48 and 6 *never* existed. Thus,

$$\frac{6 \times 8 \times 4 \times 1}{8 \times 3 \times 1 \times 1} =$$

Again, 6 in the *dividend* and 3 in the *divisor* are *divisible* by 3, since 6 divided by 3 equals 2, and 3 divided by 3 equals 1. *Cross off* the 6 and write the 2 *over* it; also, cross off the 3 and write the 1 *under* it. Thus,

$$\frac{\overset{2}{\cancel{6}} \times 8 \times 4 \times 1}{8 \times \underset{1}{\cancel{3}} \times 1 \times 1} =$$

The 6 and 3 are *not* to be considered any longer, and, in fact, may be *erased* entirely and the 2 and 1 placed in their stead, and treated as if the 6 and 3 *never* existed. Thus,

$$\frac{2 \times 8 \times 4 \times 1}{8 \times 1 \times 1 \times 1} =$$

Canceling the 8 in the dividend and the 8 in the divisor, the result is

$$\frac{2 \times \overset{1}{\cancel{8}} \times 4 \times 1}{\underset{1}{\cancel{8}} \times 1 \times 1 \times 1} = \frac{2 \times 1 \times 4 \times 1}{1 \times 1 \times 1 \times 1}.$$

Since there are *no two remaining numbers* (one in the dividend and one in the divisor) *divisible* by *any number* except 1, without a remainder, it is *impossible* to cancel further.

Multiply all the *uncanceled numbers* in the *dividend* together, and divide their *product* by the *product* of all the *uncanceled* numbers in the divisor. The *result* will be the *quotient*. The *product* of all the *uncanceled numbers* in the *dividend* equals $2 \times 1 \times 4 \times 1 = 8$; the product of all the *uncanceled* numbers in the *divisor* equals $1 \times 1 \times 1 \times 1 = 1$.

Hence, $\dfrac{2 \times 1 \times 4 \times 1}{1 \times 1 \times 1 \times 1} = \dfrac{8}{1} = 8.$ Ans.

Or, $\dfrac{\overset{2}{\cancel{6}} \times \overset{8}{\cancel{48}} \times \overset{4}{\cancel{28}} \times \overset{1}{\cancel{5}}}{\underset{\underset{1}{\cancel{8}}}{\cancel{96}} \times \underset{\underset{1}{\cancel{3}}}{\cancel{15}} \times \underset{1}{\cancel{7}} \times \underset{1}{\cancel{6}}} = \dfrac{8}{1} = 8.$ Ans.

(*b*) $(80 \times 60 \times 50 \times 16 \times 14) \div (70 \times 50 \times 24 \times 20)$.

Placing the numerator over the denominator, the problem becomes

$$\frac{80 \times 60 \times 50 \times 16 \times 14}{70 \times 50 \times 24 \times 20} = ?$$

The 50 in the *dividend* and 70 in the *divisor* are both *divisible* by 10, since 50 divided by 10 equals 5, and 70 divided by 10 equals 7. *Cross off* the 50 and write the 5 *over* it; also, *cross off* the 70 and write the 7 *under* it. Thus,

$$\frac{80 \times 60 \times \overset{5}{\cancel{50}} \times 16 \times 14}{\underset{7}{\cancel{70}} \times 50 \times 24 \times 20} =$$

The 50 and 70 are not to be considered any longer, and, in fact, may be erased entirely and the 5 and 7 placed in their stead, and treated as if the 50 and 70 *never* existed. Thus,

$$\frac{80 \times 60 \times 5 \times 16 \times 14}{7 \times 50 \times 24 \times 20} =$$

Also, 80 in the *dividend* and 20 in the *divisor* are *divisible* by 20, since 80 divided by 20 equals 4, and 20 divided by 20 equals 1. Cross off the 80 and write the 4 *over* it; also, cross off the 20 and write the 1 *under* it. Thus,

$$\frac{\overset{4}{\cancel{80}} \times 60 \times 5 \times 16 \times 14}{7 \times 50 \times 24 \times \underset{1}{\cancel{20}}} =$$

The 80 and 20 are *not* to be considered any longer, and, in fact, may be erased entirely and the 4 and 1 placed in their stead, and treated as if the 80 and 20 *never* existed. Thus,

$$\frac{4 \times 60 \times 5 \times 16 \times 14}{7 \times 50 \times 24 \times 1} =$$

Again, 16 in the *dividend* and 24 in the *divisor* are *divisible* by 8, since 16 divided by 8 equals 2, and 24 divided by 8 equals 3. *Cross off* the 16 and write the 2 *over* it; also cross off the 24 and write the 3 *under* it. Thus,

$$\frac{4 \times 60 \times 5 \times \overset{2}{\cancel{16}} \times 14}{7 \times 50 \times \underset{3}{\cancel{24}} \times 1} =$$

The 16 and 24 are not to be considered any longer, and, in fact, may be erased entirely and the 2 and 3 placed in their stead, and treated as if the 16 and 24 *never* existed. Thus,

$$\frac{4 \times 60 \times 5 \times 2 \times 14}{7 \times 50 \times 3 \times 1} =$$

Again, 60 in the *dividend* and 50 in the *divisor* are *divisible* by 10, since 60 divided by 10 equals 6, and 50 divided by 10 equals 5. *Cross off* the 60 and write the 6 *over* it; also, cross off the 50 and write the 5 *under* it. Thus,

$$\frac{4 \times \overset{6}{\cancel{60}} \times 5 \times 2 \times 14}{7 \times \underset{5}{\cancel{50}} \times 3 \times 1} =$$

The 60 and 50 are not to be considered any longer, and, in fact, may be erased entirely and the 6 and 5 placed in their stead, and treated as if the 60 and 50 *never* existed. Thus,

$$\frac{4 \times 6 \times 5 \times 2 \times 14}{7 \times 5 \times 3 \times 1} =$$

The 14 in the *dividend* and 7 in the *divisor* are *divisible* by 7, since 14 divided by 7 equals 2, and 7 divided by 7 equals 1. *Cross off* the 14 and write the 2 *over* it; also, cross off the 7 and write the 1 *under* it. Thus,

$$\frac{4 \times 6 \times 5 \times 2 \times \overset{2}{\cancel{14}}}{\underset{1}{\cancel{7}} \times 5 \times 3 \times 1} =$$

The 14 and 7 are not to be considered any longer, and, in fact, may be erased entirely and the 2 and 1 placed in their stead, and treated as if the 14 and 7 *never* existed. Thus,

$$\frac{4 \times 6 \times 5 \times 2 \times 2}{1 \times 5 \times 3 \times 1} =$$

The 5 in the *dividend* and 5 in the *divisor* are *divisible* by 5, since 5 divided by 5 equals 1. *Cross off* the 5 of the *dividend* and write the 1 *over* it; also, cross off the 5 of the *divisor* and write the 1 *under* it. Thus,

$$\frac{4 \times 6 \times \overset{1}{\cancel{5}} \times 2 \times 2}{1 \times \underset{1}{\cancel{5}} \times 3 \times 1} =$$

The 5 in the *dividend* and 5 in the *divisor* are not to be considered any longer, and, in fact, may be erased entirely and 1 and 1 placed in their stead, and treated as if the 5 and 5 *never* existed. Thus,

$$\frac{4 \times 6 \times 1 \times 2 \times 2}{1 \times 1 \times 3 \times 1} =$$

The 6 in the *dividend* and 3 in the *divisor* are *divisible* by 3, since 6 divided by 3 equals 2, and 3 divided by 3 equals 1. *Cross off* the 6 and place 2 *over* it; also, cross off the 3 and place 1 *under* it. Thus,

$$\frac{4 \times \overset{2}{\cancel{6}} \times 1 \times 2 \times 2}{1 \times 1 \times \underset{1}{\cancel{3}} \times 1} =$$

The 6 and 3 are not to be considered any longer, and, in fact, may be erased entirely and 2 and 1 placed in their stead, and treated as if the 6 and 3 *never* existed. Thus,

$$\frac{4 \times 2 \times 1 \times 2 \times 2}{1 \times 1 \times 1 \times 1} = \frac{32}{1} = 32. \quad \text{Ans.}$$

Hence, $\dfrac{\overset{4}{\cancel{80}} \times \overset{\overset{2}{\cancel{6}}}{\cancel{60}} \times \overset{\overset{1}{\cancel{5}}}{\cancel{50}} \times \overset{2}{\cancel{16}} \times \overset{2}{\cancel{14}}}{\underset{\underset{1}{7}}{\cancel{70}} \times \underset{\underset{1}{\cancel{5}}}{\cancel{50}} \times \underset{\underset{1}{\cancel{3}}}{\cancel{24}} \times \underset{1}{\cancel{20}}} = \dfrac{4 \times 2 \times 1 \times 2 \times 2}{1 \times 1 \times 1 \times 1} = \dfrac{32}{1} = 32.$ Ans.

(19) 28 acres of land at \$133 an acre would cost 28 × \$133 = \$3,724.

$$\begin{array}{r} 2\,8 \\ \hline 1\,0\,6\,4 \\ 2\,6\,6 \\ \hline \$3\,7\,2\,4 \end{array}$$

If a mechanic earns $1,500 a year and his expenses are $968 per year, then he would save $1500—$968, or $532 per year.

$$\begin{array}{r} 9\,6\,8 \\ \hline \$5\,3\,2 \end{array}$$

If he saves $532 in 1 year, to save $3,724 it would take as many years as $532 is contained times in $3,724, or 7 years.

$$5\,3\,2\,)\,3\,7\,2\,4\,(\,7 \text{ years.   Ans.}$$
$$\underline{3\,7\,2\,4}$$

(**20**) If the freight train ran 365 miles in one week, and 3 times as far lacking 246 miles the next week, then it ran (3 × 365 miles) — 246 miles, or 849 miles the second week. Thus,

$$\begin{array}{r} 3\,6\,5 \\ 3 \\ \hline 1\,0\,9\,5 \\ 2\,4\,6 \\ \hline \textit{difference}\quad 8\,4\,9 \end{array}$$ miles. Ans.

(**21**) The distance from Philadelphia to Pittsburg is 354 miles. Since there are 5,280 feet in one mile, in 354 miles there are 354 × 5,280 feet, or 1,869,120 feet. If the driving wheel of the locomotive is 16 feet in circumference, then in going from Philadelphia to Pittsburg, a distance of 1,869,-120 feet, it will make 1,869,120 ÷ 16, or 116,820 revolutions.

$$\begin{array}{r} 16\,)\,1\,8\,6\,9\,1\,2\,0\,(\,1\,1\,6\,8\,2\,0 \text{ rev.   Ans.} \\ 1\,6 \\ \hline 2\,6 \\ 1\,6 \\ \hline 1\,0\,9 \\ 9\,6 \\ \hline 1\,3\,1 \\ 1\,2\,8 \\ \hline 3\,2 \\ 3\,2 \\ \hline 0 \end{array}$$

(22)   (a)   5 7 6) 5 8 9 8 2 4 (1 0 2 4   Ans.

```
          5 7 6
          ─────
          1 3 8 2
          1 1 5 2
          ───────
            2 3 0 4
            2 3 0 4
            ───────
```

(b)   4 3 9 1 1) 3 6 9 7 3 0 6 2 0 (8 4 2 0   Ans.

```
            3 5 1 2 8 8
            ───────────
              1 8 4 4 2 6
              1 7 5 6 4 4
              ───────────
                  8 7 8 2 2
                  8 7 8 2 2
                  ─────────
                          0
```

(c)   5 0 5) 2 5 2 7 5 2 5 (5 0 0 5   Ans.

```
          2 5 2 5
          ───────
            2 5 2 5
            2 5 2 5
            ───────
```

(d)   1 2 3 4) 4 9 6 1 7 9 4 3 0 2 (4 0 2 0 9 0 3   Ans

```
          4 9 3 6
          ───────
            2 5 7 9
            2 4 6 8
            ───────
              1 1 1 4 3
              1 1 1 0 6
              ─────────
                  3 7 0 2
                  3 7 0 2
                  ───────
```

(23)   The harness evidently cost the difference between
$444 and the amount which he paid for the horse and wagon.
Since $264 + $153 = $417, the amount paid for the horse
and wagon, $444 − $417 = $27, the cost of the harness.

```
      $2 6 4                    $4 4 4
       1 5 3                     4 1 7
       ─────                     ─────
      $4 1 7                    $2 7   Ans.
```

(24) *(a)*

$$
\begin{array}{r}
1024 \\
576 \\
\hline
6144 \\
7168 \\
5120 \\
\hline
589824 \quad \text{Ans.}
\end{array}
$$

(b)

$$
\begin{array}{r}
5005 \\
505 \\
\hline
25025 \\
250250 \\
\hline
2527525 \quad \text{Ans.}
\end{array}
$$

(c)

$$
\begin{array}{r}
43911 \\
8420 \\
\hline
878220 \\
175644 \\
351288 \\
\hline
369730620 \quad \text{Ans.}
\end{array}
$$

(25) Since there are 12 months in a year, the number of days the man works is $25 \times 12 = 300$ days. As he works 10 hours each day, the number of hours that he works in one year is $300 \times 10 = 3,000$ hours. Hence, he receives for his work $3,000 \times 30 = 90,000$ cents, or $90,000 \div 100 = \$900$. Ans.

(26) See Art. **71.**

(27) See Art. **77.**

(28) See Art. **73.**

(29) See Art. **73.**

(30) See Art. **75.**

(31) $\frac{13}{8}$ is an improper fraction, since its numerator 13 is greater than its denominator 8.

(32) $4\frac{1}{2}$; $14\frac{3}{10}$; $85\frac{4}{19}$.

(33) To reduce a fraction to its lowest terms means to change its form without changing its value. In order to do this, we must divide both numerator and denominator by the same number until we can no longer find any number (except 1) which will divide both of these terms without a remainder.

To reduce the fraction $\frac{4}{8}$ to its lowest terms we divide both numerator and denominator by 4, and obtain as a result the fraction $\frac{1}{2}$. Thus, $\frac{4 \div 4}{8 \div 4} = \frac{1}{2}$; similarly, $\frac{4 \div 4}{16 \div 4} = \frac{1}{4}$; $\frac{8 \div 4}{32 \div 4} = \frac{2 \div 2}{8 \div 2} = \frac{1}{4}$; $\frac{32 \div 8}{64 \div 8} = \frac{4 \div 4}{8 \div 4} = \frac{1}{2}$. Ans.

(34) When the denominator of any number is not expressed, it is understood to be 1, so that $\frac{6}{1}$ is the same as $6 \div 1$, or 6. To reduce $\frac{6}{1}$ to an improper fraction whose denominator is 4, we must multiply both numerator and denominator by some number which will make the denominator of 6 equal to 4. Since this denominator is 1, by multiplying both terms of $\frac{6}{1}$ by 4 we shall have $\frac{6 \times 4}{1 \times 4} = \frac{24}{4}$, which has the *same value* as 6, but has a *different form*. Ans.

(35) In order to reduce a mixed number to an improper fraction, we must *multiply the whole number by the denominator of the fraction* and *add the numerator of the fraction to that product*. This *result* is the *numerator of the improper fraction*, of which the *denominator is the denominator of the fractional part of the mixed number*.

$7\frac{7}{8}$ means the same as $7 + \frac{7}{8}$. In 1 there are $\frac{8}{8}$, hence in 7 there are $7 \times \frac{8}{8} = \frac{56}{8}$; $\frac{56}{8}$ plus the $\frac{7}{8}$ of the mixed number $= \frac{56}{8} + \frac{7}{8} = \frac{63}{8}$, which is the required improper fraction.

$13\frac{5}{16} = \frac{(13 \times 16) + 5}{16} = \frac{213}{16}$; $10\frac{3}{4} = \frac{(10 \times 4) + 3}{4} = \frac{43}{4}$.

(36) The value of a fraction is obtained by dividing the numerator by the denominator.

To obtain the value of the fraction $\frac{13}{2}$ we divide the numerator 13 by the denominator 2. 2 is contained in 13 six times, with 1 remaining. This 1 remaining is written over the denominator 2, thereby making the fraction $\frac{1}{2}$, which is annexed to the whole number 6, and we obtain $6\frac{1}{2}$ as the mixed number. The reason for performing this operation is the following: In 1 there are $\frac{2}{2}$ (two halves), and in $\frac{13}{2}$ (thirteen halves) there are as many units (1) as 2 is contained times in 13, which is 6, and $\frac{1}{2}$ (one-half) unit remaining.

Hence, $\frac{13}{12} = 6 + \frac{1}{2} = 6\frac{1}{2}$, the required mixed number. Ans.

$\frac{17}{4} = 4\frac{1}{4}$. Ans. $\frac{69}{16} = 4\frac{5}{16}$. Ans. $\frac{16}{8} = 2$. Ans. $\frac{67}{64} = 1\frac{3}{64}$. Ans.

(37) In division of fractions, *invert the divisor* (or, in other words, turn it upside down) *and proceed as in multiplication.*

(a) $35 \div \frac{5}{16} = \frac{35}{1} \times \frac{16}{5} = \frac{35 \times 16}{1 \times 5} = \frac{560}{5} = 112$. Ans.

(b) $\frac{9}{16} \div 3 = \frac{9}{16} \div \frac{3}{1} = \frac{9}{16} \times \frac{1}{3} = \frac{9 \times 1}{16 \times 3} = \frac{9}{48} = \frac{3}{16}$. Ans.

(c) $\frac{17}{2} \div 9 = \frac{17}{2} \div \frac{9}{1} = \frac{17}{2} \times \frac{1}{9} = \frac{17 \times 1}{2 \times 9} = \frac{17}{18}$. Ans.

(d) $\frac{113}{64} \div \frac{7}{16} = \frac{113}{64} \times \frac{16}{7} = \frac{113 \times 16}{64 \times 7} = \frac{1,808}{448} = \frac{452}{112} = \frac{113}{28}$)$113(4\frac{1}{28}$. Ans.

$\underline{112}$

1

(c) $15\frac{3}{4} \div 4\frac{3}{8} = ?$ Before proceeding with the division, reduce both of the mixed numbers to improper fractions. Thus, $15\frac{3}{4} = \frac{(15 \times 4) + 3}{4} = \frac{60 + 3}{4} = \frac{63}{4}$, and $4\frac{3}{8} = \frac{(4 \times 8) + 3}{8} = \frac{32 + 3}{8} = \frac{35}{8}$. The problem is now $\frac{63}{4} \div \frac{35}{8} = ?$ As before, invert the divisor and multiply; $\frac{63}{4} \div \frac{35}{8} = \frac{63}{4} \times \frac{8}{35} = \frac{63 \times 8}{4 \times 35} = \frac{504}{140} = \frac{252}{70} = \frac{126}{35} = \frac{18}{5}$.

$$\frac{18}{5}) \, 18 \, (3\frac{3}{5} \quad \text{Ans.}$$
$$\frac{15}{3}$$

(38) $\quad \frac{1}{8} + \frac{2}{8} + \frac{5}{8} = \frac{1 + 2 + 5}{8} = \frac{8}{8} = 1.$ Ans.

When the *denominators* of the fractions to be added *are alike*, we know that the units are divided into the *same number of parts* (in this case *eighths*); we, therefore, *add the numerators* of the fractions to find the number of parts (eighths) taken or considered, thereby obtaining $\frac{8}{8}$ or 1 as the sum.

(39) When the *denominators* are *not alike* we know that the units are divided into *unequal parts*, so before adding them we must find a common denominator for the denominators of all the fractions. Reduce the fractions to fractions having this common denominator, add the numerators and write the sum over the common denominator.

In this case, the least common denominator, or the least number that will contain all the denominators, is 16; hence, we must reduce all these fractions to sixteenths and then add their numerators.

$\frac{1}{4} + \frac{3}{8} + \frac{5}{16} = ?$ To reduce the fraction $\frac{1}{4}$ to a fraction having 16 for a denominator, we must multiply both terms

of the fraction by some number which will make the denom-
inator 16. This number evidently is 4, hence, $\frac{1 \times 4}{4 \times 4} = \frac{4}{16}$.

Similarly, both terms of the fraction $\frac{3}{8}$ must be multiplied
by 2 to make the denominator 16, and we have $\frac{3 \times 2}{8 \times 2} = \frac{6}{16}$.
The fractions now have a common denominator 16; hence,
we find their sum by adding the numerators and placing their
sum over the common denominator, thus: $\frac{4}{16} + \frac{6}{16} + \frac{5}{16} =$
$\frac{4 + 6 + 5}{16} = \frac{15}{16}$. Ans.

(40) When mixed numbers and whole numbers are to be
added, add the fractional parts of the mixed numbers separately, and if the resulting fraction is an improper fraction,
reduce it to a whole or mixed number. Next, add all the
whole numbers, including the one obtained from the addition
of the fractional parts, and annex to their sum the fraction
of the mixed number obtained from reducing the improper
fraction.

$42 + 31\frac{5}{8} + 9\frac{7}{16} = ?$ Reducing $\frac{5}{8}$ to a fraction having
a denominator of 16, we have $\frac{5}{8} \times \frac{2}{2} = \frac{10}{16}$. Adding the two
fractional parts of the mixed numbers we have $\frac{10}{16} + \frac{7}{16} =$
$\frac{10 + 7}{16} = \frac{17}{16} = 1\frac{1}{16}$.

The problem now becomes $42 + 31 + 9 + 1\frac{1}{16} = ?$

42
31
9
$1\frac{1}{16}$

$83\frac{1}{16}$ Ans.

Adding all the whole numbers and the
number obtained from adding the fractional
parts of the mixed numbers, we obtain $83\frac{1}{16}$
as their sum.

(41) $29\frac{3}{4} + 50\frac{5}{8} + 41 + 69\frac{3}{16} = ?$ $\quad \frac{3}{4} = \frac{3 \times 4}{4 \times 4} = \frac{12}{16}.$

$\frac{5}{8} = \frac{5 \times 2}{8 \times 2} = \frac{10}{16}.$ $\quad \frac{12}{16} + \frac{10}{16} + \frac{3}{16} = \frac{12 + 10 + 3}{16} = \frac{25}{16} = 1\frac{9}{16}.$

The problem now becomes $29 + 50 + 41 + 69 + 1\frac{9}{16} = ?$

$$\begin{array}{ll} 29 & \text{square inches.} \\ 50 & \text{square inches.} \\ 41 & \text{square inches.} \\ 69 & \text{square inches.} \\ \underline{1\frac{9}{16}} & \text{square inches.} \\ 190\frac{9}{16} & \text{square inches. \quad Ans.} \end{array}$$

(42) $(a)\ \dfrac{7}{\dfrac{3}{16}} = 7 \div \frac{3}{16} = 7 \times \frac{16}{3} = \frac{7 \times 16}{3} = \frac{112}{3} = 37\frac{1}{3}.$ Ans.

The line between 7 and $\frac{3}{16}$ means that 7 is to be divided by $\frac{3}{16}.$

$(b)\ \dfrac{\dfrac{15}{32}}{\dfrac{5}{8}} = \frac{15}{32} \div \frac{5}{8} = \frac{\overset{3}{\cancel{15}}}{\cancel{32}} \times \frac{\cancel{8}}{\cancel{5}} = \frac{\cancel{15} \times \cancel{8}}{\cancel{32} \times \cancel{5}} = \frac{3}{4}.$ Ans.

$(c)\ \dfrac{\dfrac{4+3}{2+6}}{5} = \dfrac{\dfrac{7}{8}}{5} = \frac{7}{8 \times 5} = \frac{7}{40}.$ (See Art. **131.**) Ans.

(43) $\frac{7}{8} = $ value of the fraction, and $28 = $ the numerator. We find that 4 multiplied by $7 = 28$, so multiplying 8, the denominator of the fraction, by 4, we have 32 for the required denominator, and $\frac{28}{32} = \frac{7}{8}.$ Hence, 32 is the required denominator. Ans.

(44) $(a)\ \frac{7}{8} - \frac{7}{16} = ?$ When the *denominators* of fractions are *not alike* it is evident that the units are divided into *unequal parts*, therefore, before subtracting, *reduce the* *fractions to fractions having a common denominator.* Then, *subtract the numerators, and place the remainder over the common denominator.*

$$\frac{7 \times 2}{8 \times 2} = \frac{14}{16}. \quad \frac{14}{16} - \frac{7}{16} = \frac{14 - 7}{16} = \frac{7}{16}. \quad \text{Ans.}$$

(*b*) $13 - 7\frac{7}{16} = ?$ This problem may be solved in two ways:

First: $13 = 12\frac{16}{16}$, since $\frac{16}{16} = 1$, and $12\frac{16}{16} = 12 + \frac{16}{16} = 12 + 1 = 13$.

$12\frac{16}{16}$ We can now subtract the whole numbers sepa-
$7\frac{7}{16}$ rately, and the fractions separately, and obtain $12 - 7$
$\overline{5\frac{9}{16}}$ $= 5$ and $\frac{16}{16} - \frac{7}{16} = \frac{16 - 7}{16} = \frac{9}{16}.$ $5 + \frac{9}{16} = 5\frac{9}{16}.$ Ans.

Second: By reducing both numbers to improper fractions having a denominator of 16.

$$13 = \frac{13}{1} = \frac{13 \times 16}{1 \times 16} = \frac{208}{16}. \quad 7\frac{7}{16} = \frac{(7 \times 16) + 7}{16} = \frac{112 + 7}{16} = \frac{119}{16}.$$

Subtracting, we have $\dfrac{208}{16} - \dfrac{119}{16} = \dfrac{208 - 119}{16} = \dfrac{89}{16}$ and

$$\frac{89}{16} = 16)\overline{89}(5\frac{9}{16}$$
$$\underline{80}$$
$$\overline{9}$$
$$\underline{16}$$

the same result that was obtained by the first method.

(*c*) $312\frac{9}{16} - 229\frac{5}{32} = ?$ We first reduce the fractions of the two mixed numbers to fractions having a common denominator. Doing this we have $\dfrac{9}{16} = \dfrac{9 \times 2}{16 \times 2} = \dfrac{18}{32}.$ We can now subtract the whole numbers and fractions separately, and have $312 - 229 = 83$ and $\dfrac{18}{32} - \dfrac{5}{32} = \dfrac{18 - 5}{32} = \dfrac{13}{32}.$

$$312\frac{18}{32}$$
$$229\frac{5}{32}$$
$$\overline{83\frac{13}{32}} \quad 83 + \frac{13}{32} = 83\frac{13}{32}. \quad \text{Ans.}$$

(45) The man evidently traveled $85\frac{5}{12} + 78\frac{9}{15} + 125\frac{17}{35}$ miles.

Adding the fractions separately in this case,

$$\frac{5}{12} + \frac{9}{15} + \frac{17}{35} = \frac{5}{12} + \frac{3}{5} + \frac{17}{35} = \frac{175 + 252 + 204}{420} = \frac{631}{420} = 1\frac{211}{420}.$$

Adding the whole numbers and the mixed number representing the sum of the fractions, the sum is $289\frac{211}{420}$ miles.   Ans.

To find the least common denominator, we have

$$
\begin{array}{l}
85 \\
78 \\
125 \\
1\frac{211}{420} \\
\hline
289\frac{211}{420}
\end{array}
$$

$$
\begin{array}{r}
5\,)\,12,\ 5,\ 35 \\
7\,)\,\overline{12,\ 1,\ \ 7} \\
\hline
12,\ 1,\ \ 1, \text{ or } 5\times7\times12 = 420.
\end{array}
$$

(46) $573\frac{4}{5}$ tons. $\frac{4}{5} = \frac{32}{40}$

$216\frac{5}{8}$ tons. $\frac{5}{8} = \frac{25}{40}$

difference $357\frac{7}{40}$ tons.   Ans. $\frac{7}{40} = difference.$

(47) Reducing $9\frac{1}{4}$ to an improper fraction, it becomes $\frac{37}{4}$. Multiplying $\frac{37}{4}$ by $\frac{3}{8}$, $\frac{37}{4} \times \frac{3}{8} = \frac{111}{32} = 3\frac{15}{32}$ dollars. Ans

(48) Referring to Arts. **114** and **116**,

$\frac{2}{3}$ of $\frac{3}{4}$ of $\frac{7}{11}$ of $\frac{19}{20}$ of 11 multiplied by $\frac{7}{8}$ of $\frac{5}{6}$ of $45 =$

$$\frac{2\times3\times7\times19\times11\times7\times5\times\overset{\overset{3}{15}}{45}}{\underset{4}{3}\times4\times11\times\underset{}{20}\times1\times8\times\underset{2}{6}\times1} = \frac{7\times19\times7\times5\times3}{4\times4\times8} = \frac{13,965}{128} =$$

$109\frac{13}{128}.$ Ans.

(49) $\frac{3}{4}$ of $16 = \frac{3}{4}\times\frac{\overset{4}{16}}{1} = 12.$ $12 \div \frac{2}{3} = \frac{\overset{6}{12}}{1}\times\frac{3}{2} = 18.$ Ans.

(50) $211\frac{1}{4}\times1\frac{7}{8} = \frac{845}{4}\times\frac{15}{8}$, reducing the mixed numbers to improper fractions. $\dfrac{845}{4} \times \dfrac{15}{8} = \dfrac{12,675}{32}$ cents $=$ amount paid for the lead. The number of pounds sold is evidently

$$\dfrac{12,675}{32} \div 2\dfrac{1}{2} = \dfrac{\overset{2,535}{\cancel{12,675}}}{\underset{16}{\cancel{32}}} \times \dfrac{\cancel{2}}{5} = \dfrac{2,535}{16} = 158\dfrac{7}{16} \text{ pounds.}$$ The

amount remaining is $211\dfrac{1}{4} - 158\dfrac{7}{16} = \dfrac{845}{4} - \dfrac{2,535}{16} = \dfrac{3,380}{16} -$

$\dfrac{2,535}{16} = \dfrac{845}{16} = 52\dfrac{13}{16}$ pounds. Ans.

(51) $\cdot\ \underset{\text{tenths.}}{0}\ \underset{\text{hundredths.}}{8} = $ *Eight **hundredths**.*

$.\underset{\text{tenths.}}{1}\ \underset{\text{hundredths.}}{3}\ \underset{\text{thousandths.}}{1} = $ *One hundred thirty-one **thousandths**.*

$\cdot\ \underset{\text{tenths.}}{0}\ \underset{\text{hundredths.}}{0}\ \underset{\text{thousandths.}}{0}\ \underset{\text{ten-thousandths.}}{1} = $ *One **ten-thousandth**.*

$\underset{\text{tenths.}}{0}\ \underset{\text{hundredths.}}{0}\ \underset{\text{thousandths.}}{0}\ \underset{\text{ten-thousandths.}}{0}\ \underset{\text{hundred-thousandths.}}{2}\ \underset{\text{millionths.}}{7} = $ *Twenty-seven **millionths**.*

$\underset{\text{tenths.}}{0}\ \underset{\text{hundredths.}}{1}\ \underset{\text{thousandths.}}{0}\ \underset{\text{ten-thousandths.}}{8} = $ *One hundred eight **ten-thousandths**.*

tenths. hundredths. thousandths. ten-thousandths.

93.0 1 0 1 = Ninety-three, and *one hundred one ten-thousandths*

In reading decimals, read the number just as you would if there were no ciphers before it. Then count from the decimal point towards the right, beginning with tenths, to as many places as there are figures, and the *name* of the last figure must be annexed to the previous reading of the figures to give the decimal reading. Thus, in the first example above, the simple reading of the figure is *eight*, and the name of its position in the decimal scale is **hundredths,** so that the decimal reading is *eight* **hundredths.** Similarly, the figures in the fourth example are ordinarily read *twenty-seven ;* the name of the position of the figure 7 in the decimal scale is **millionths,** giving, therefore, the decimal reading as *twenty-seven* **millionths.**

If there should be a whole number before the decimal point, read it as you would read any whole number, and read the decimal as you would if the whole number were not there; or, read the whole number and then say, "and" so many hundredths, thousandths, or whatever it may be, as "ninety-three, *and* one hundred one ten thousandths."

(52) See Art. **139.**

(53) See Art. **153.**

(54) See Art. **160.**

(55) A fraction is one or more of the equal parts of a unit, and is expressed by a numerator and a denominator, while a decimal fraction is a number of *tenths, hundredths, thousandths*, etc., of a unit, and is expressed by placing a period (.), called a decimal point, to the left of the figures of the number, and omitting the denominator.

(56) See Art. **165.**

(**57**) To reduce the fraction $\frac{1}{2}$ to a decimal, we annex one cipher to the numerator, which makes it 1.0. Dividing 1.0, the numerator, by 2, the denominator, gives a quotient of .5, the decimal point being placed before the *one* figure of the quotient, or .5, since only *one* cipher was annexed to the numerator. Ans.

$$\begin{array}{r} 7 \\ \overline{8\,)\,7.0\,0\,0} \\ \hline .8\,7\,5 \quad \text{Ans.} \end{array}$$

Since $.65 = \frac{65}{100}$, then, $\frac{65}{100}$ must equal .65. Or, when the denominator is 10, 100, 1000, etc., point off as many places in the numerator as there are ciphers in the denominator. Doing so, $\frac{65}{100} = .65$. Ans.

$$\begin{array}{r} 5 \\ \overline{32\,)\,5.0\,0\,0\,0\,0}\,(.1\,5\,6\,2\,5 \quad \text{Ans.} \\ 3\,2 \\ \hline 1\,8\,0 \\ 1\,6\,0 \\ \hline 2\,0\,0 \\ 1\,9\,2 \\ \hline 8\,0 \\ 6\,4 \\ \hline 1\,6\,0 \\ 1\,6\,0 \\ \hline \end{array}$$

$\frac{125}{1000} = .125.$ Ans.

(**58**) (*a*) This example, written in the form of a fraction, means that the numerator $(32.5 + .29 + 1.5)$ is to be divided by the denominator $(4.7 + 9)$. The operation is as follows:

$$\frac{32.5 + .29 + 1.5}{4.7 + 9} = ?$$

$$\begin{array}{r} 3\,2.5 \\ + \quad .2\,9 \\ + \quad 1.5 \\ \hline \end{array}$$

$$\begin{array}{r} 1\,3.7\,)\,3\,4.2\,9\,0\,0\,0\,(\,2.5\,0\,2\,9 \quad \text{Ans.} \\ 2\,7\,4 \\ \hline 6\,8\,9 \\ 6\,8\,5 \\ \hline 4\,0\,0 \\ 2\,7\,4 \\ \hline 1\,2\,6\,0 \\ 1\,2\,3\,3 \\ \hline 2\,7 \end{array}$$

$$\begin{array}{r} 4.7 \\ + \quad 9.0 \\ \hline 1\,3.7 \end{array}$$

Since there are 5 decimal places in the dividend and 1 in the divisor, there are $5 - 1$ or 4 places to be pointed off in the quotient. The fifth figure of the decimal is evidently less than 5, (*b*) Here again the problem is to divide the numerator, which is $(1.283 \times \overline{8+5})$, by the denominator, which is 2.63. The operation is as follows:

$$\frac{1.283 \times \overline{8+5}}{2.63} = ? \quad \overline{8+5} = 13.$$

$$
\begin{array}{r}
1.283 \\
\times\quad 13 \\
\hline
3849 \\
1283 \\
\end{array}
$$

$$2.63)\overline{16.679000}(6.3418 \quad \text{Ans.}$$

$$
\begin{array}{r}
1578 \\
\hline
899 \\
789 \\
\hline
1100 \\
1052 \\
\hline
480 \\
\end{array}
\qquad
\begin{array}{r}
480 \\
263 \\
\hline
2170 \\
2104 \\
\hline
66 \\
\end{array}
$$

(*c*) $\dfrac{\overline{589+27} \times \overline{163-8}}{25+39} = ?$

$$
\begin{array}{r}
163 \\
-\quad 8 \\
\hline
155 \\
\times\,616 \\
\hline
930 \\
155 \\
930 \\
\end{array}
$$

$$
\begin{array}{r}
589 \\
+\ 27 \\
\hline
616 \\
\end{array}
$$

$$64)\overline{95480.000}(1491.875$$

$$\text{Ans.}$$

$$
\begin{array}{r}
64 \\
\hline
314 \\
256 \\
\hline
588 \\
576 \\
\hline
120 \\
64 \\
\hline
560 \\
512 \\
\hline
480 \\
448 \\
\hline
320 \\
320 \\
\hline
\end{array}
$$

$$
\begin{array}{r}
25 \\
+39 \\
\hline
64 \\
\end{array}
$$

There are three decimal places in the quotient, since three ciphers were annexed to the dividend.

(*d*) $\dfrac{\overline{40.6 + 7.1} \times (3.029 - 1.874)}{6.27 + 8.53 - 8.01} = ?$

$$\begin{array}{r} 40.6 \\ + 7.1 \\ \hline 47.7 \end{array} \qquad \begin{array}{r} 3.02\,9 \\ -\ 1.87\,4 \\ \hline 1.15\,5 \end{array}$$

$$\begin{array}{r} 6.2\,7 \\ +\ 8.5\,3 \\ \hline 14.80 \\ -\ 8.01 \\ \hline 6.7\,9 \end{array} \qquad \begin{array}{r} \times\ \ 47.7 \\ \hline 8\,0\,8\,5 \\ 8\,0\,8\,5 \\ 4\,6\,2\,0 \\ \hline \end{array}$$

6 decimal places in the dividend − 2 decimal places in the divisor = 4 decimal places to be pointed off in the quotient.

$6.79\,)\,5\,5.0\,9\,3\,5\,0\,0\,(\,8.1\,1\,3\,9.\quad$ Ans

$$\begin{array}{r} 5\,4\,3\,2 \\ \hline 7\,7\,3, \\ 6\,7\,9 \\ \hline 9\,4\,5 \\ 6\,7\,9 \\ \hline 2\,6\,6\,0 \\ 2\,0\,3\,7 \\ \hline 6\,2\,3\,0 \\ 6\,1\,1\,1 \\ \hline 1\,1\,9 \end{array}$$

(59) $\quad .875 = \dfrac{875}{1,000} = \dfrac{175}{200} = \dfrac{7}{8}$ of a foot.

1 foot = 12 inches.

$\dfrac{7}{8}$ of 1 foot $= \dfrac{7}{\overset{}{\underset{2}{\cancel{8}}}} \times \dfrac{\overset{3}{\cancel{12}}}{1} = \dfrac{21}{2} = 10\dfrac{1}{2}$ inches.   Ans.

(60)   12 inches = 1 foot.

$\dfrac{3}{16}$ of an inch $= \dfrac{3}{16} \div 12 = \dfrac{\cancel{3}}{16} \times \dfrac{1}{\underset{4}{\cancel{12}}} = \dfrac{1}{64}$ of a foot.

Point off 6 decimal places in the quotient, since we annexed six ciphers to the dividend, the divisor containing no decimal places; hence, 6 − 0 = 6 places to be pointed off.

$$\frac{1}{64}) 1.000000 (.015625 \text{ Ans.}$$

```
        64
       ----
       3 6 0
       3 2 0
       -----
         4 0 0
         3 8 4
         -----
           1 6 0
           1 2 8
           -----
             3 2 0
             3 2 0
             -----
```

(61) If 1 cubic inch of water weighs .03617 of a pound, the weight of 1,500 cubic inches will be .03617 × 1,500 = 54.255 lb.

```
    .0 3 6 1 7  lb.
        1 5 0 0
    -------------
    1 8 0 8 5 0 0
    3 6 1 7
    -------------
    5 4.2 5 5 0 0  lb.   Ans.
```

(62) 72.6 feet of fencing at $.50 a foot would cost

$$72.6 \times .50, \text{ or } \$36.30.$$

```
    7 2.6 × .50, or $36.30.
        .5 0
    -----------
    $3 6.3 0 0
```

If, by selling a carload of coal at a profit of $1.65 per ton, I make $36.30, then there must be as many tons of coal in the car as 1.65 is contained times in 36.30, or 22 tons.

```
    1.6 5 ) 3 6.3 0 ( 2 2 tons.   Ans.
          3 3 0
          -----
            3 3 0
            3 3 0
            -----
```

(63) $231\,)\,17892.00000\,(\,77.45454$, or 77.4545 to
 four decimal places. Ans.

$$
\begin{array}{r}
1617 \\ \hline
1722 \\
1617 \\ \hline
1050 \\
924 \\ \hline
1260 \\
1155 \\ \hline
1050 \\
924 \\ \hline
1260 \\
1155 \\ \hline
1050
\end{array}
$$

(64)

$$\dfrac{\cancel{74.26}\times\cancel{24}\times\cancel{3.1416}\times 19\times 19\times 350}{\underset{1,000}{\cancel{33,000}}\times\cancel{12}\times\underset{2}{\cancel{4}}} =$$

with $37.13 \quad \cancel{2} \quad .0952$

$$\dfrac{37.13\times .0952\times 19\times 19\times 350}{1,000}=\dfrac{446{,}618.947600}{1,000}= \cdot$$

446.619 to three decimal places. Ans.

$$
\begin{array}{cccc}
37.13 & 19 & 361 & 3.534776 \\
.0952 & 19 & 350 & 126350 \\ \hline
7426 & 171 & 18050 & 176738800 \\
18565 & 19 & 1083 & 10604328 \\
33417 & \overline{361} & \overline{126350} & 21208656 \\ \hline
3.534776 & & & 7069552 \\
& & & 3534776 \\ \hline
& & & 446618.947600
\end{array}
$$

(65) See Art. **174.** Applying rule in Art. **175,**

(a) $.7928\times\dfrac{64}{64}=\dfrac{50.7392}{64}=\dfrac{51}{64}.$ Ans.

(b) $.1416\times\dfrac{32}{32}=\dfrac{4.5312}{32}=\dfrac{5}{32}.$ Ans.

(c) $.47915\times\dfrac{16}{16}=\dfrac{7.6664}{16}=\dfrac{8}{16}=\dfrac{1}{2}.$ Ans.

(66) In subtraction of decimals, *place the decimal points directly under each other*, and proceed as in the subtraction of whole numbers, placing the *decimal point in the remainder directly under the decimal points above.*

$$(a) \quad \begin{array}{r} 709.6300 \\ .8514 \\ \hline 708.7786 \end{array} \text{ Ans.}$$

In the above example we proceed as follows: We can not subtract 4 ten-thousandths from 0 ten-thousandths, and, as there are no thousandths, we take 1 hundredth from the three hundredths. 1 *hundredth* = 10 *thousandths* = 100 *ten-thousandths*. 4 ten-thousandths from 100 ten-thousandths leaves 96 ten-thousandths. 96 ten-thousandths = 9 *thousandths* + 6 *ten-thousandths*. Write the 6 ten-thousandths in the ten-thousandths place in the remainder. The next figure in the subtrahend is 1 thousandth. This must be subtracted from the 9 thousandths which is a part of the 1 hundredth taken previously from the 3 hundredths. Subtracting, we have 1 thousandth from 9 thousandths leaves 8 thousandths, the 8 being written in its place in the remainder. Next we have to subtract 5 hundredths from 2 hundredths (1 hundredth having been taken from the 3 hundredths makes it but 2 hundredths now). Since we can not do this, we take 1 tenth from 6 tenths. 1 tenth (= 10 hundredths) + 2 hundredths = 12 hundredths. 5 hundredths from 12 hundredths leaves 7 hundredths. Write the 7 in the hundredths place in the remainder. Next we have to subtract 8 tenths from 5 tenths (5 tenths now, because 1 tenth was taken from the 6 tenths). Since this can not be done, we take 1 unit from the 9 units. 1 *unit* = 10 *tenths;* 10 tenths + 5 tenths = 15 tenths, and 8 tenths from 15 tenths leaves 7 tenths. Write the 7 in the tenths place in the remainder. In the minuend we now have 708 units (one unit having been taken away) and 0 units in the subtrahend. 0 units from 708 units leaves 708 units; hence, we write 708 in the remainder.

$$(b) \quad \begin{array}{r} 81.963 \\ 1.700 \\ \hline 80.263 \end{array} \text{ Ans.} \qquad (c) \quad \begin{array}{r} 18.00 \\ .18 \\ \hline 17.82 \end{array} \text{ Ans.} \qquad (d) \quad \begin{array}{r} 1.000 \\ .001 \\ \hline .999 \end{array} \text{ Ans.}$$

(*e*) 872.1 − (.8721 + .008) = ? In this problem we are to subtract (.8721 + .008) from
872.1. First perform the operation as indicated by the sign between the decimals
enclosed by the parenthesis.

.8721
.008
―――
.8801 *sum.*

Subtracting the sum (obtained by adding the decimals

872.1000
.8801
―――――
871.2199 Ans.

enclosed within the parenthesis) from
the number 872.1 (as required by the
minus sign before the parenthesis),
we obtain the required remainder.

(*f*) (5.028 + .0073) − (6.704 − 2.38) = ? First perform
the operations as indicated by the signs between the numbers enclosed by the parentheses. The first parenthesis shows that
5.028 and .0073 are to be added. This
gives 5.0353 as their sum.

5.0280
.0073
―――
5.0353 *sum.*

6.704
2.380
―――
4.324 *difference.*

The second parenthesis shows that
2.38 is to be subtracted from 6.704.
The difference is found to be 4.324.

The sign between the parentheses indicates that the

5.0353
4.324
―――
.7113 Ans.

quantities obtained by performing
the above operations, are to be subtracted, namely, that 4.324 is to be
subtracted from 5.0353. Perform-
ing this operation we obtain .7113 as the final result.

(**67**) In subtracting a decimal from a fraction, or subtracting a fraction from a decimal, either reduce the fraction
to a decimal before subtracting, or reduce the decimal to a
fraction and then subtract.

(*a*) $\frac{7}{8}$ − .807 = ? $\frac{7}{8}$ reduced to a decimal becomes

$$\frac{7}{8})7.000$$
.875

.875
.807
―――
.068 Ans.

Subtracting .807 from .875 the remainder is .068, as shown.

(*b*) $.875 - \dfrac{3}{8} = ?$ Reducing .875 to a fraction we have

$.875 = \dfrac{875}{1,000} = \dfrac{175}{200} = \dfrac{35}{40} = \dfrac{7}{8}$; hence, $\dfrac{7}{8} - \dfrac{3}{8} = \dfrac{7-3}{8} = \dfrac{4}{8} = \dfrac{1}{2}.$

Ans.

Or, by reducing $\dfrac{3}{8}$ to a decimal, $\dfrac{3}{8}$) 3.000 and then sub-

.3 7 5

tracting, we obtain $.875 - .375 = .5 = \dfrac{5}{10} =$ .8 7 5

.3 7 5

$\dfrac{1}{2}$, the same answer as above.   .5 0 0 Ans.

(*c*) $\left(\dfrac{5}{32} + .435\right) - \left(\dfrac{21}{100} - .07\right) = ?$ We first perform the operations as indicated by the signs between the numbers enclosed by the parentheses. Reduce $\dfrac{5}{32}$ to a decimal and we obtain $\dfrac{5}{32} = .15625$ (see example 7).

Adding .15625 and .435,   .1 5 6 2 5 $\dfrac{21}{100} = .21$; subtracting, .2 1

.4 3 5   .0 7

sum .5 9 1 2 5   *difference* .1 4

We are now prepared to perform the   .5 9 1 2 5
operation indicated by the minus sign be-   .1 4
tween the parentheses, which is, *difference* .4 5 1 2 5 Ans.

(*d*) This problem means that 33 millionths and 17 thousandths are to be added. Also, that 53 hundredths and 274 thousandths are to be added, and the smaller of these sums is to be subtracted from the larger sum. Thus, $(.53 + .274) - (.000033 + .017) = ?$

.000033
.017
.017033 *sum.*

.53
.274
.804 *sum.*

.804   *larger sum,*
.017033 *smaller sum,*
difference .786967 Ans.

(68) In addition of decimals the *decimal points must be placed directly under each other*, so that *tenths* will come *under tenths, hundredths under hundredths, thousandths under thousandths*, etc. The addition is then performed as in whole numbers, *the decimal point of the sum being placed directly under the decimal points above.*

```
.125
.7
.089
.4005
.9
.000027
2.214527   Ans.
```

(69)
```
927.416
  8.274
372.6
 62.07938
1370.36938  Ans.
```

(70)
```
tenths.
hundredths.
thousandths.
ten-thousandths.
hundred-thousandths.
millionths.
.017
.2
.000047
.217047 = Ans.
```

(71) (*a*) There are 3 decimal places in the multiplicand and 3 in the multiplier; hence, there are 3 + 3 or 6 decimal places in the product. Since the product contains but four figures, we prefix two ciphers in order to obtain the necessary six decimal places.

```
.107
.013
 321
107
.001391  Ans.
```

There are two decimal places in the multiplier and none in the multiplicand; hence, there are 2 + 0 or two decimal places in the first product.

Since there are 2 decimal places in the multiplicand and 3 decimal places in the multiplier, there are 3 + 2 or 5 decimal places in the second product.

(*b*)
```
203
2.03
609
4060
412.09
.203
123627
824180
83.65427  Ans.
```

(*c*) First perform the operations indicated by the signs between the numbers enclosed by the parenthesis, and then perform whatever may be required by the sign before the parenthesis.

Multiply together the numbers 2.7 and 31.85.

The parenthesis shows that .316 is to be taken from 3.16.

```
3.160
 .316
_____
2.844
```

```
 31.85
   2.7
_____
 22295
 6370
_____
85.995
```

The product obtained by the first operation is now multiplied by the remainder obtained by performing the operation indicated by the signs within the parenthesis.

```
 85.995
  2.844
_____
 343980
 343980
 687960
171990
_____
244.569780  Ans
```

(*d*) $(107.8 + 6.541 - 31.96) \times 1.742 = ?$

```
    107.8
+   6.541
_____
  114.341
-  31.96
_____
   82.381
×   1.742
_____
   164762
   329524
   576667
   82381
_____
143.507702  Ans.
```

(**72**) (*a*) $\left(\frac{7}{16} - .13\right) \times \overline{.625 + \frac{5}{8}} = ?$

First perform the operation indicated by the parenthesis.

$$\frac{7}{16} = \frac{7}{16}) 7.0000 (.4375$$

$$
\begin{array}{r}
64 \\
\hline
60 \\
48 \\
\hline
120 \\
112 \\
\hline
80 \\
80 \\
\hline
\end{array}
$$

We point off four decimal places since we annexed four ciphers.

$$
\begin{array}{r}
.4375 \\
.13 \\
\hline
\end{array}
$$

Subtracting, we obtain   .3075

The vinculum has the same meaning as the parenthesis;

$$\frac{5}{8} = \frac{5}{8}) 5.000$$
.625

hence, we perform the operation indicated by it. We point off three decimal places, since three ciphers were annexed to the 5.

Adding the terms included by the vinculum, we obtain

$$
\begin{array}{r}
.625 \\
.625 \\
\hline
1.250 \\
\end{array}
$$

The final operation is to perform the work indicated by the sign between the parenthesis and the vinculum, thus,

$$
\begin{array}{r}
.3075 \\
1.25 \\
\hline
15375 \\
6150 \\
3075 \\
\hline
.384375 \quad \text{Ans.}
\end{array}
$$

(b) $\left(\frac{19}{32} \times .21\right) - \left(.02 \times \frac{3}{16}\right) = ?$

$.21 = \frac{21}{100}.$ $\frac{19}{32} \times \frac{21}{100} = \frac{399}{3200}.$ $.02 = \frac{2}{100}.$ $\frac{2}{100} \times \frac{3}{16} = \frac{6}{1600} = \frac{3}{800}.$

$\frac{3}{800} = \frac{3 \times 4}{800 \times 4} = \frac{12}{3200}.$ $\frac{399}{3200} - \frac{12}{3200} = \frac{399 - 12}{3200} = \frac{387}{3200}.$

Reducing $\frac{387}{3200}$ to a decimal, we obtain

$\frac{387}{3200}$) 3 8 7.0 0 0 0 0 0 0 (.1 2 0 9 3 7 5   Ans.

3 2 0 0
_____
6 7 0 0
6 4 0 0
_____
3 0 0 0 0
2 8 8 0 0   Point off seven decimal
_____   places, since seven ciphers
1 2 0 0 0   were annexed to the divi-
9 6 0 0   dend.
_____
2 4 0 0 0
2 2 4 0 0
_____
1 6 0 0 0
1 6 0 0 0

(c) $\left(\dfrac{13}{4}+.013-2.17\right)\times\overline{13\dfrac{1}{4}-7\dfrac{5}{16}}=?$

$\dfrac{13}{4}=\dfrac{13}{4}$) 1 3.0 0   Point off two decimal   3.2 5
        3.2 5   places, since two ciphers   + .0 1 3
                were annexed to the divi-   _____
                dend.   3.2 6 3

$\dfrac{5}{16}$ reduced to a decimal is .3125, since   − 2.1 7
                                                            _____
                                                             1.0 9 3

$.\overline{16}$) 5.0 0 0 0 (.3 1 2 5   Point off four decimal
        4 8   places, since four ciphers
        ___   were annexed to the
        2 0   dividend.
        1 6
        ___
        4 0
        3 2
        ___
        8 0
        8 0

Then, $7\dfrac{5}{16}=7.3125$, and $13\dfrac{1}{4}=13.25$, since $\dfrac{1}{4}=\dfrac{1}{4}$) 1.0 0
                                                                            _____
                                                                              .2 5

$$\begin{array}{r} 1\,3.2\,5 \\ -\ \ 7.3\,1\,2\,5 \\ \hline 5.9\,3\,7\,5 \end{array}$$

$$\begin{array}{r} 5.9\,3\,7\,5 \\ \times\ \ 1.0\,9\,3 \\ \hline 1\,7\,8\,1\,2\,5 \\ 5\,3\,4\,3\,7\,5 \\ 5\,9\,3\,7\,5\,0 \\ \hline 6.4\,8\,9\,6\,8\,7\,5 \ \ \text{Ans.} \end{array}$$

(73) (a) $.875 \div \frac{1}{2} = .875 \div .5 \left(\text{since } \frac{1}{2} = .5\right) = 1.75.$ Ans.
Another way of solving this is to reduce $.875$ to its equivalent common fraction and then divide.

$.875 = \frac{7}{8}$, since $.875 = \frac{875}{1,000} = \frac{175}{200} = \frac{35}{40} = \frac{7}{8}$; then, $\frac{7}{8} \div$

$\frac{1}{2} = \frac{7}{\underset{4}{8}} \times \frac{2}{1} = \frac{7}{4} = 1\frac{3}{4}.$ Since $\frac{3}{4} = 4) 3.00 (.75, \quad 1\frac{3}{4} = 1.75,$

the same answer as above.
$$\begin{array}{r} 2\,8 \\ \hline 2\,0 \\ 2\,0 \end{array}$$

(b) $\frac{7}{8} \div .5 = \frac{7}{8} \div \frac{1}{2} \left(\text{since } .5 = \frac{1}{2}\right) = \frac{7}{\underset{4}{8}} \times \frac{2}{1} = \frac{7}{4} = 1\frac{3}{4}$, or
1.75. Ans.

This can also be solved by reducing $\frac{7}{8}$ to its equivalent decimal and dividing by $.5$; $\frac{7}{8} = .875$; $.875 \div .5 = 1.75.$ Since there are three decimal places in the dividend and one in the divisor, there are $3-1$, or 2 decimal places in the quotient.

(c) $\frac{.375 \times \frac{1}{4}}{\frac{5}{16} - .125} = ?$ We shall solve this problem by first reducing the decimals to their equivalent common fractions.

$.375 = \frac{375}{1,000} = \frac{75}{200} = \frac{15}{40} = \frac{3}{8}.$ $\frac{3}{8} \times \frac{1}{4} = \frac{3}{32}$, or the value of the numerator of the fraction.

$.125 = \frac{125}{1,000} = \frac{25}{200} = \frac{1}{8}.$ Reducing $\frac{1}{8}$ to sixteenths, we

have $\frac{1 \times 2}{8 \times 2} = \frac{2}{16}.$ Then, $\frac{5}{16} - \frac{2}{16} = \frac{3}{16}$, or the value of the denominator of the fraction. The problem is now reduced to

$$\frac{\frac{3}{32}}{\frac{3}{16}} = ? \qquad \frac{\frac{3}{32}}{\frac{3}{16}} = \frac{3}{32} \div \frac{3}{16} = \frac{\cancel{3}}{\cancel{32}} \times \frac{\cancel{16}}{\cancel{3}} = \frac{1}{2} \text{ or } .5. \quad \text{Ans.}$$

(**74**) $\dfrac{1.25 \times 20 \times 3}{\dfrac{87 + (11 \times 8)}{459 + 32}} = ?$ In this problem $1.25 \times 20 \times 3$ constitutes the numerator of the complex fraction.

$$\begin{array}{r} 1.2\,5 \\ \times \quad 2\,0 \\ \hline 2\,5.0\,0 \\ \times \quad 3 \\ \hline 7\,5 \end{array}$$ Multiplying the factors of the numerator together, we find their product to be 75.

The fraction $\dfrac{87 + (11 \times 8)}{459 + 32}$ constitutes the denominator of the complex fraction. The value of the numerator of this fraction equals $87 + 88 = 175$.

The numerator is combined as though it were written $87 + (11 \times 8)$, and its result is

$$\begin{array}{r} 1\,1 \\ 8 \text{ times} \\ \hline 8\,8 \\ +\,8\,7 \\ \hline 1\,7\,5 \end{array}$$

The value of the denominator of this fraction is equal to $459 + 32 = 491$. The problem then becomes

$$\frac{75}{\frac{175}{491}} = \frac{75}{1} \div \frac{175}{491} = \frac{75}{1} \times \frac{491}{175} = \frac{\overset{3}{\cancel{75}} \times 491}{\underset{7}{\cancel{175}}} = \frac{1,473}{7} = 210\frac{3}{7}. \quad \text{Ans.}$$

(**75**) 1 plus $.001 = 1.001$. $.01$ plus $.000001 = .010001$. And $1.001 - .010001 =$

$$\begin{array}{r} 1.0\,0\,1 \\ .0\,1\,0\,0\,0\,1 \\ \hline .9\,9\,0\,9\,9\,9 \quad \text{Ans.} \end{array}$$

ARITHMETIC.

(76) A certain per cent. of a number means so many hundredths of that number.

25% of 8,428 lb. means 25 hundredths of 8,428 lb. Hence, 25% of 8,428 lb. = .25 × 8,428 lb. = 2,107 lb. Ans.

(77) Here $100 is the base and 1% = .01 is the rate. Then, .01 × $100 = $1. Ans.

(78) $\frac{1}{2}$% means one-half of one per cent. Since 1% is .01, $\frac{1}{2}$% is .005, for, $2\overline{)\,.0\,1\,0}$ And .005 × $35,000 = $175.
$\quad\quad\quad\quad\quad\quad\quad\quad\quad\quad\quad .0\,0\,5$ Ans.

(79) Here 50 is the base, 2 is the percentage; and it is required to find the rate. Applying rule, Art. **193,**

$$\text{rate} = \text{percentage} \div \text{base};$$
$$\text{rate} = 2 \div 50 = .04 \text{ or } 4\%. \quad \text{Ans.}$$

(80) By Art. **193,** rate = percentage ÷ base.*

As percentage = 10 and base = 10, we have rate = 10 ÷ 10 = 1 = 100%. Hence, 10 is 100% of 10. Ans.

(81) (a) Rate = percentage ÷ by base. Art. **193.**

As percentage = $176.54 and base = $2,522, we have

$$\text{rate} = 176.54 \div 2,522 = .07 = 7\%. \quad \text{Ans.}$$

$$2\,5\,2\,2\,\overline{)\,1\,7\,6.5\,4}$$
$$.0\,7$$

---

* Remember that an expression of this form means that the first term is to be *divided by* the second term. Thus, as above, it means percentage *divided by* base.

(*b*) Base = percentage ÷ rate.   Art. **192.**

As percentage = 16.96 and rate = 8% = .08, we have

$$\text{base} = 16.96 \div .08 = 212. \quad \text{Ans.}$$

$$.08\,)\,\underline{1\,6.\,9\,6}$$
$$2\,1\,2$$

(*c*) Amount is the sum of the base and percentage; hence, the percentage = amount minus the base.

Amount = 216.7025 and base = 213.5; hence, percentage = 216.7025 − 213.5 = 3.2025.

Rate = percentage ÷ base.   Art. **193.**

Therefore, rate = 3.2025 ÷ 213.5 = .015 = 1½%.   Ans.

$$2\,1\,3.5\,)\,3.2\,0\,2\,5\,(\,.0\,1\,5 = 1\tfrac{1}{2}\%$$
$$\underline{2\,1\,3\,5}$$
$$\underline{1\,0\,6\,7\,5}$$
$$1\,0\,6\,7\,5$$

(*d*) The difference is the remainder found by subtracting the percentage from the base; hence, base − the difference = the percentage.   Base = 207 and difference = 201.825, hence percentage = 207 − 201.825 = 5.175.

Rate = percentage ÷ base.   Art. **193.**

Therefore, rate = 5.175 ÷ 207 = .025 = $.02\frac{1}{2}$ = $2\frac{1}{2}$%.   Ans.

$$2\,0\,7\,)\,5.1\,7\,5\,(\,.0\,2\,5$$
$$\underline{4\,1\,4}$$
$$\underline{1\,0\,3\,5}$$
$$1\,0\,3\,5$$

(**82**) In this problem $5,500 is the amount, since it equals what he paid for the farm + what he gained; 15% is the rate, and the cost (to be found) is the base. Applying rule, Art. **197,**

$$\text{base} = \text{amount} \div (1 + \text{rate}); \text{ hence,}$$
$$\text{base} = \$5,500 \div (1 + .15) = \$4,782.61. \quad \text{Ans.}$$

```
1.15 ) 5 5 0 0.0 0 0 0 ( 4 7 8 2.61
       4 6 6
       ─────
       9 0 0
       8 0 5
         ─────
         9 5 0
         9 2 0
           ─────
           ,3 0 0
           2 3 0
             ─────
             7 0 0
             6 9 0
               ─────
               1 0 0
               1 1 5
```

The example can also be solved as follows: 100% = cost; if he gained 15%, then 100 + 15 = 115% = $5,500, the selling price.

If 115% = $5,500, 1% = $\frac{1}{115}$ of $5,500 = $47.8261, and 100%, or the cost, = 100 × $47.8261 = $4,782.61. Ans.

(83) 24 % of $950 = .24 × 950 = $228

$12\frac{1}{2}$% of $950 = .125 × 950 = 118.75

17 % of $950 = .17 × 950 = 161.50

$53\frac{1}{2}$% of $950   = $508.25

The total amount of his yearly expenses, then, is $508.25, hence his savings are $950 − $508.25 = $441.75. Ans.

Or, as above, 24% + $12\frac{1}{2}$% + 17% = $53\frac{1}{2}$%, the total percentage of expenditures; hence, 100% − $53\frac{1}{2}$% = $46\frac{1}{2}$% = per cent. saved. And $950 × .465 = $441.75 = his yearly savings. Ans.

(84) The percentage is 961.38, and the rate is $37\frac{1}{2}$. By Art. 192,

Base = percentage ÷ rate
= 961.38 ÷ .375 = 2,563.68, the number. Ans.

Another method of solving is the following:

If $37\frac{1}{2}$% of a number is 961.38, then $.37\frac{1}{2}$ times the number = 961.38 and the number = $961.38 \div .37\frac{1}{2}$, which, as above = 2,563.68. Ans.

```
.3 7 5 ) 9 6 1.3 8 0 0 0 ( 2 5 6 3.6 8
         7 5 0
         ─────
         2 1 1 3
         1 8 7 5
         ───────
           2 3 8 8
           2 2 5 0
           ───────
             1 3 8 0
             1 1 2 5
             ───────
               2 5 5 0
               2 2 5 0
               ───────
                 3 0 0 0
                 3 0 0 0
                 ───────
```

(85) Here \$1,125 is 30% of some number; hence, \$1,125 = the percentage, 30% = the rate, and the required number is the base. Applying rule, Art. **192,**

Base = percentage ÷ rate = \$1,125 ÷ .30 = \$3,750.

Since \$3,750 is $\frac{3}{4}$ of the property, one of the fourths is $\frac{1}{3}$ of \$3,750 = \$1,250, and $\frac{4}{4}$ or the entire property, is 4 × \$1,250 = \$5,000. Ans.

(86) Here \$4,810 is the difference and 35% the rate. By Art. **198,**

Base = difference ÷ (1 − rate)
= \$4,810 ÷ (1 − .35) = \$4,810 ÷ .65 = \$7,400. Ans.

```
.6 5 ) 4 8 1 0.0 0 ( 7 4 0 0
       4 5 5
       ─────
         2 6 0            1.0 0
         2 6 0             .3 5
         ─────            ─────
           0 0             .6 5
```

Solution can also be effected as follows: 100% = the sum diminished by 35%, then (1 − .35) = .65, which is \$4,810.

If $65\% = \$4,810$, $1\% = \dfrac{1}{65}$ of $4,810 = \$74$, and $100\% = 100 \times \$74 = \$7,400$. Ans.

(**87**) In this example the sales on Monday amounted to $\$197.55$, which was $12\frac{1}{2}\%$ of the sales for the entire week; i. e., we have given the percentage, $\$197.55$, and the rate, $12\frac{1}{2}\%$, and the required number (or the amount of sales for the week) equals the base. By Art. **192**,

$$\text{Base} = \text{percentage} \div \text{rate} = \$197.55 \div .125;$$

or,
```
.125)197.5500(1580.4  Ans.
     125
     ───
     725
     625
     ───
     1005
     1000
     ────
      500
      500
      ───
```

Therefore, base $= \$1,580.40$, which also equals the sales for the week.

(**88**) 16.5 miles $= 12\frac{1}{2}\%$ of the entire length of the road. We wish to find the *entire* length.

16.5 miles is the percentage, $12\frac{1}{2}\%$ is the rate, and the entire length will be the base. By Art. **192**,

$$\text{Base} = \text{percentage} \div \text{rate} = 16.5 \div .12\frac{1}{2}.$$

```
.125)16.500(132 miles.  Ans.
     125
     ───
     400
     375
     ───
      250
      250
      ───
```

(89) Here we have given the difference, or $35, and the rate, or 60%, to find the base. We use the rule in Art. **198,**

Base = difference ÷ (1 − rate)

= $35 ÷ (1 − .60) = $35 ÷ .40 = $87.50. Ans.

$$.40\,)\,3\,5.0\,0\,0\,(\,8\,7.5$$
$$3\,2\,0$$
$$\overline{3\,0\,0}$$
$$2\,8\,0$$
$$\overline{2\,0\,0}$$
$$2\,0\,0$$

Or, 100% = whole debt; 100% − 60% = 40% = $35.

If 40% = $35, then 1% = $\frac{1}{40}$ of $35 = $\frac{35}{40}$, and 100% = $\frac{35}{40} \times 100 = 87.50. Ans.

(90) 28 rd. 4 yd. 2 ft. 10 in. to inches.

$$\times \quad 5\tfrac{1}{2}$$
$$\overline{1\,5\,4}$$
$$+\quad 4$$
$$\overline{1\,5\,8}\ \text{yards}$$
$$\times \quad 3$$
$$\overline{4\,7\,4}$$
$$+\quad 2$$
$$\overline{4\,7\,6}\ \text{feet}$$
$$\times \quad 1\,2$$
$$\overline{5\,7\,1\,2}$$
$$+\quad 1\,0$$
$$\overline{5\,7\,2\,2}\ \text{inches. Ans.}$$

Since there are 5½ yards in one rod, in 28 rods there are 28 × 5½ or 154 yards; 154 yards plus 4 yards = 158 yards. There are 3 feet in one yard; therefore, in 158 yards there are 3 × 158 or 474 feet; 474 feet + 2 feet = 476 feet. There are 12 inches in one foot, and in 476 feet there are 12 × 476 or 5,712 inches; 5,712 inches + 10 inches = 5,722 inches. Ans.

(91)

12) 5 7 2 2 inches.

3) 4 7 6 + 10 inches.

5½) 1 5 8 + 2 feet.

2 8 + 4 yards.

Ans. = 28 rd. 4 yd. 2 ft. 10 in.

EXPLANATION.—There are 12 inches in 1 foot; hence, in 5,722 inches there are as many feet as 12 is contained times in 5,722 inches, or 476 ft. and 10 inches remaining. Write these 10 inches as a remainder. There are 3 feet in 1 yard; hence, in 476 feet there are as many yards as 3 is contained times in 476 feet, or 158 yards and 2 feet remaining. There are $5\frac{1}{2}$ yards in one rod; hence, in 158 yards there are 28 rods and 4 yards remaining. Then, in 5,722 inches there are 28 rd. 4 yd. 2 ft. 10 in.

(92)   5   weeks 3.5 days.
        × 7
        ———
        3 5   days in 5 weeks.
        + 3.5
        ———
        3 8.5 days.

Then, we find how many seconds there are in 38.5 days.

        3 8.5 days
        × 2 4 hours in one day.
        ———
        1 5 4 0
        7 7 0
        ———
        9 2 4.0 hours in 38.5 days.
        × 6 0 minutes in one hour.
        ———
        5 5 4 4 0 minutes in 38.5 days.
        × 6 0 seconds in one minute.
        ———
        3 3 2 6 4 0 0 seconds in 38.5 days.   Ans.

(93)   Since there are 24 gr. in 1 pwt., in 13,750 gr. there are as many pennyweights as 24 is contained times in 13,750, or 572 pwt. and 22 gr. remaining. Since there are 20 pwt. in 1 oz., in 572 pwt. there are as many ounces as 20 is contained times in 572, or 28 oz. and 12 pwt. remaining.

Since there are 12 oz. in 1 lb. (Troy), in 28 oz. there are as many pounds as 12 is contained times in 28, or 2 lb. and 4 oz. remaining. We now have the pounds and ounces required by the problem; therefore, in 13,750 gr. there are 2 lb. 4 oz. 12 pwt. 22 gr.

24) 1 3 7 5 0 gr.
   20) 5 7 2 pwt. + 22 **gr.**
      12) 2 8 oz. + 12 pwt.
            2 lb. + 4 oz.

Ans. = 2 lb. 4 oz. 12 pwt. 22 gr.

(94)   100) 4 7 6 3 2 5 4 li.
                     80) 4 7 6 3 2 + 54 li.
                          5 9 5 + 32 ch.

Ans. = 595 mi. 32 ch. 54 li.

EXPLANATION.—There are 100 li. in one chain; hence, in 4,763,254 li. there are as many chains as 100 is contained times in 4,763,254 li., or 47,632 ch. and 54 li. remaining. Write the 54 li. as a remainder. There are 80 ch. in one mile; hence, in 47,632 ch. there are as many miles as 80 is contained times in 47,632 ch., or 595 miles and 32 ch. remaining.

Then, in 4,763,254 li. there are 595 mi. 32 ch. 54 li.

(95)   1728) 7 6 4 3 2 5 cu. in.
               27) 4 4 2 + 549 cu. in.
                    1 6 cu. yd. + 10 cu. ft.

Ans. = 16 cu. yd. 10 cu. ft. 549 cu. in.

EXPLANATION.—There are 1,728 cu. in. in one cubic foot; hence, in 764,325 cu. in. there are as many cubic feet as 1,728 is contained times in 764,325, or 442 cu. ft. and 549 cu. in. remaining. Write the 549 cu. in. as a remainder. There are 27 cu. ft. in one cubic yard; hence, in 442 cu. ft. there are as many cubic yards as 27 is contained times in 442 cu. ft., or 16 cu. yd. and 10 cu. ft. remaining. Then, in 764,325 cu. in. there are 16 cu. yd. 10 cu. ft. 549 cu. in.

(96) We must arrange the different terms in columns, taking care to have like denominations in the same column.

	rd.	yd.	ft.	in.
	2	2	2	3
		4	1	9
			2	7
	3	2½	0	7
or	3	2	2	1	Ans.

ARITHMETIC.

Explanation.—We begin to add at the right-hand column. $7 + 9 + 3 = 19$ in.; as 12 in. make one foot, 19 in. = 1 ft. and 7 in. Place the 7 in. in the inches column, and reserve the 1 ft. to add to the next column.

1 (reserved) $+ 2 + 1 + 2 = 6$ ft. Since 3 ft. make 1 yard, 6 ft. $= 2$ yd. and 0 ft. remaining. Place the cipher in the column of feet and reserve the 2 yd. for the next column.

2 (reserved) $+ 4 + 2 = 8$ yd. Since $5\frac{1}{2}$ yd. $= 1$ rod, 8 yd. = 1 rd. and $2\frac{1}{2}$ yd. Place $2\frac{1}{2}$ yd. in the yards column, and reserve 1 rd. for the next column; 1 (reserved) $+ 2 = 3$ rd.

Ans. = 3 rd. $2\frac{1}{2}$ yd.   0 ft.   7 in.
or,   3 rd.   2 yd.   1 ft.   13 in.
or,   3 rd.   2 yd.   2 ft.   1 in. Ans.

(97) We write the compound numbers so that the units of the same denomination shall stand in the same column. Beginning to add with the lowest denomination, we find that the sum of the gills is $1 + 2 + 3 = 6$. Since there are 4 gi. in 1 pint, in 6 gi. there are as many pints as 4 is contained times in 6, or 1 pt. and 2 gi. We place 2 gi. under the gills column and reserve the 1 pt. for the pints column; the sum of the

gal.	qt.	pt.	gi.
3	3	1	3
6	0	1	2
4	0	0	1
8	5	0	

16 gal. 3 qt. 0 pt. 2 gi.

pints is 1 (reserved) $+ 5 + 1 + 1 = 8$. Since there are 2 pt. in 1 quart, in 8 pt. there are as many quarts as 2 is contained times in 8, or 4 qt. and 0 pt. We place the cipher under the column of pints and reserve the 4 for the quarts column. The sum of the quarts is 4 (reserved) $+ 8 + 3 = 15$. Since there are 4 qt. in 1 gallon, in 15 qt. there are as many gallons as 4 is contained times in 15, or 3 gal. and 3 qt. remaining. We now place the 3 under the quarts column and reserve the 3 gal. for the gallons column. The sum of the gallons column is 3 (reserved) $+ 4 + 6 + 3 = 16$ gal. Since we can not reduce 16 gal. to any higher denomination, we have 16 gal. 3 qt. 0 pt. and 2 gi. for the answer.

(98) Reduce the grains, pennyweights, and ounces to higher denominations.

24) 240 gr.   20) 125 pwt.   12) 50 oz.

  10 pwt.   6 oz. 5 pwt.   4 lb. 2 oz.

Then, 3 lb. + 4 lb. 2 oz. + 6 oz. 5 pwt. + 10 pwt. =

lb.	oz.	pwt.
3		
4	2	
	6	5
		10
7 lb.	8 oz.	15 pwt. Ans.

(99) Since "seconds" is the lowest denomination in this problem, we find their sum first, which is 11 + 29 + 25 + 30 + 12, or 107 seconds. Since there are 60 seconds in 1 minute, in 107″ there are as many minutes as 60 is contained times in 107, or 1 minute and 47 seconds remaining. We place the 47 under the seconds column and reserve the 1 for the minutes column. The sum of the minutes is 1 (reserved) +

deg.	min.	sec.
11	16	12
13	19	30
20	0	25
0	26	29
10	17	11
55°	19′	47″

17 + 26 + 19 + 16, or 79. Since there are 60 minutes in 1 degree, in 79 minutes there are as many degrees as 60 is contained times in 79, or 1 degree and 19 minutes remaining. We place the 19 under the minutes column and reserve the 1 degree for the degrees column. The sum of the degrees is 1 (reserved) + 10 + 20 + 13 + 11, or 55 degrees. Since we can not reduce 55 degrees to any higher denominations, we have 55° 19′ 47″ for the answer.

(100) Since "inches" is the lowest denomination in this problem, we find their sum first, which is 11 + 8 + 6, or 25 inches. Since there are 12 inches in 1 foot, in 25 inches there are as many feet as 12 is contained times in 25, or 2 feet and 1 inch remaining. Place the 1 inch under the inches column, and reserve the 2 feet to add to the column of feet. The sum of the feet is 2 feet (reserved) $+ 2 + 1 =$ 5 feet. Since there are 3 feet in 1 yard, in 5 feet there are as many yards as 3 is contained times in 5 feet, or 1 yard and 2 feet remaining. Place the 2 feet under the column of feet, and reserve the 1 yard to add to the column of yards. The sum of

	rd.	yd.	ft.	in.
	130	5	1	6
	215	0	2	8
	304	4	0	11
	650	$4\frac{1}{2}$	2	1
mi.				
or, 2	10	5	0	7 Ans.

the yards is 1 yard (reserved) $+ 4 + 5 = 10$ yards. Since there are $5\frac{1}{2}$ yards in 1 rod, in 10 yards there are as many rods as $5\frac{1}{2}$ is contained times in 10, or 1 rod and $4\frac{1}{2}$ yards remaining. Place the $4\frac{1}{2}$ yards under the column of yards, and reserve the 1 rod for the column of rods. The sum of the rods is 1 (reserved) $+ 304 + 215 + 130 = 650$ rods. Place 650 rods under the column of rods. Therefore, the sum is 650 rd. $4\frac{1}{2}$ yd. 2 ft. 1 in. Or, since $\frac{1}{2}$ a yard $= 1$ ft. 6 in., and since there are 320 rods in 1 mile; the sum may be expressed as 2 mi. 10 rd. 5 yd. 0 ft. 7 in. Ans.

(101) Since "square links" is the lowest denomination in this problem, we find their sum first, which is $21 + 23 + 16 + 18 + 23 + 21$, or 122 square links. Place 122 square links under the column of square links. The sum of the square rods is $2 + 3 + 2 + 2 + 2 + 3$, or 14 square rods. Place 14 square rods under the column of square rods. The sum of the square chains

A.	sq. ch.	sq. rd.	sq. li.
21	67	3	21
28	78	2	23
47	6	2	18
56	59	2	16
25	38	3	23
46	75	2	21
255	3	14	122

is 323 square chains. Since there are 10 square chains in 1 acre, in 323 square chains there are as many acres as 10 is contained times in 323 square chains, or 32 acres and 3 square chains remaining. Place 3 square chains under the column of square chains, and reserve the 32 acres to add to the column of acres. The sum of the acres is 32 acres (reserved) + 46 + 25 + 56 + 47 + 28 + 21, or 255 acres. Place 255 acres under the column of acres. Therefore, the sum is 255 A. 3 sq. ch. 14 sq. rd. 122 sq. li. Ans.

(**102**) Before we can subtract 300 ft. from 20 rd. 2 yd. 2 ft. and 9 in., we must reduce the 300 ft. to higher denominations.

Since there are 3 feet in 1 yard, in 300 feet there are as many yards as 3 is contained times in 300, or 100 yards. There are $5\frac{1}{2}$ yards in 1 rod, hence in 100 yards there are as many rods as $5\frac{1}{2}$ or $\frac{11}{2}$ is contained times in $100 = 18\frac{2}{11}$ rods.

$$100 \div \frac{11}{2} = 100 \times \frac{2}{11} = \frac{100 \times 2}{11} = \frac{200}{11}$$

$$\frac{200}{11}) \, 2\,0\,0 \, (\, 18\frac{2}{11} \text{ rd.}$$
$$\underline{1\,1}$$
$$9\,0$$
$$\underline{8\,8}$$
$$2$$

Since there are $5\frac{1}{2}$ or $\frac{11}{2}$ yards in 1 rod, in $\frac{2}{11}$ rods there are $\frac{2}{11} \times \frac{11}{2}$, or one yard, so we find that 300 feet equals 18 rods and 1 yard. The problem now is as follows: From 20 rd. 2 yd. 2 ft. and 9 in. take 18 rd. and 1 yd.

We place the smaller number under the larger one, so that units of the same denomination fall in the same column. Beginning with the lowest denomination, we see that 0 inches from 9 inches leaves 9 inches. Going to the next higher denomination, we see that 0 feet from 2 feet leaves 2 feet. Subtracting 1 yard from 2

rd.	yd.	ft.	in.
20	2	2	9
18	1	0	0
2	1	2	9

yards, we have 1 yard remaining, and 18 rods from 20 rods leaves 2 rods. Therefore, the difference is 2 rd. 1 yd. 2 ft. 9 in. Ans.

(103)

A.	sq. rd.	sq. yd.
114	80	25
75	70	30
39	9	$25\frac{1}{4}$ Ans.

EXPLANATION.—Place the subtrahend under the minuend so that like denominations are under each other. Then begin at the right with the lowest denomination. We can not subtract 30 from 25, so we take one square rod $(= 30\frac{1}{4}$ square yards) from 80 square rods, leaving 79 square rods; adding $30\frac{1}{4}$ square yards to 25 square yards, we have $55\frac{1}{4}$ square yards; subtracting 30 from $55\frac{1}{4}$ square yards leaves $25\frac{1}{4}$ square yards; we now subtract 70 square rods from 79 square rods, which leaves 9 square rods; next, we subtract 75 acres from 114 acres, which leaves 39 acres, which we place under the column of acres.

(104) If 10 gal. 2 qt. and 1 pt. of molasses are sold from a hogshead at one time, and 26 gal. 3 qt. are sold at another time, then the total amount of molasses sold equals 10 gal. 2 qt. 1 pt. plus 26 gal. 3 qt.

Since the pint is the lowest denomination, we add the pints first, which equal $0 + 1$, or 1 pint. We can not reduce 1 pint to any higher denomination, so we place it under the pint column. The number of quarts is $3 + 2$, or 5. Since there are 4 quarts in 1 gallon, in 5 quarts there are as many gallons as 4 is contained times in 5, or 1 gallon and 1 quart remaining. We place the 1 quart under the quart column, and reserve the 1 gallon to add to the column of

gal.	qt.	pt.
10	2	1
26	3	0
37 gal.	1 qt.	1 pt.

gallons. The number of gallons equals 1 (reserved) + 26 + 10, or 37 gallons.

If 37 gal. 1 qt. and 1 pt. are sold from a hogshead of molasses (63 gal.), there remains the difference between 63 gal. and 37 gal. 1 qt. 1 pt., or 25 gal. 2 qt. and 1 pt.

63 gal. is the same as 62 gal. 3 qt. 2 pt., since 1 gal. equals 4 qt. and 1 qt. = 2 pt.

Beginning with the lowest denomination, 1 pt. from the

gal.	qt.	pt.
62	3	2
37	1	1
25	2	1

2 pt. 1 pint from 2 pints leaves 1 pint. One quart from 3 quarts leaves 2 quarts, and 37 gallons from 62 gallons leaves 25 gallons. Therefore, there are 25 gal. 2 qt. and 1 pt. of molasses remaining in the hogshead. Ans.

(**105**) If a person were born June 19, 1850, in order to find how old he would be on Aug. 3, 1892, subtract the earlier date from the later date.

On August 3, 7 mo. and 3 da. have elapsed from the beginning of the year, and on June 19, 5 mo. and 19 da.

Beginning with the lowest denomination, we find that 19 days can not be taken from 3 days, so we take 1 month from 7 months. The 1 month which we took equals 30 days, for

yr.	mo.	da.
1892	7	3
1850	5	19
42	1	14

in all cases 30 days are allowed to a month. Adding 30 days to the 3 days, we have 33 days; subtracting 19 days from 33 days, we have 14 days remaining. Since we borrowed 1 month from the months column, we have 7 − 1, or 6 months remaining; subtracting 5 months from 6 months, we have 1 month remaining. 1850 from 1892 leaves 42 years. Therefore, he would be 42 years 1 month and 14 days old. Ans.

(**106**) If a note given Aug. 5, 1890, were paid June 3, 1892, in order to find the length of time it was due, subtract the earlier date from the later date.

Beginning with the lowest denomination, we find that 5 can not be subtracted from 3, so we take a unit from the next

yr.	mo.	da.
1892	5	3
1890	7	5
1	9	28

higher denomination, which is months. The 1 month which we take equals 30 days. Adding the 30 days to the 3 days, we have 33 days. 5 days from 33 days leaves 28 days. Since we took 1 month from the months column, only 4 months remain. 7 months cannot be taken from 4 months, so we take 1 year from the years column, which equals 12 months. 12 months + 4 months = 16 months. 7 months from 16 months = 9 months. Since we took 1 year from the years column, we have 1892 − 1, or 1891 remaining. 1890 from 1891 leaves 1 year. Hence, the note ran 1 year 9 months and 28 days. Ans.

(107) Write the number of the year, month, day, hour, and minute of the earlier date under the year, month, day, hour, and minute of the later date, and subtract.

22 minutes before 8 o'clock is the same as 38 minutes after 7 o'clock. 7 o'clock P. M. is 19 hours from the beginning of the day, as there are 12 hours in the morning and 7 in the afternoon. December is 11 months from the beginning of the year.

10 o'clock A. M. is 10 hours from the beginning of the day. July is 6 months from the beginning of the year. The minuend would be the later date, or 1,888 years, 11 months, 11 days, 19 hours, and 38 minutes.

The subtrahend would be the earlier date, or 1,883 years, 6 months, 3 days, 10 hours, and 16 minutes.

Subtracting, we have

yr.	mo.	da.	hr.	min.
1888	11	11	19	38
1883	6	3	10	16
5	5	8	9	22

or, 5 yr. 5 mo. 8 da. 9 hr. and 22 min. Ans.

16 minutes subtracted from 38 minutes leaves 22 minutes; 10 hours from 19 hours leaves 9 hours; 3 days from 11 days leaves 8 days; 6 months subtracted from 11 months leaves 5 months; 1,883 from 1,888 leaves 5 years.

(**108**) In multiplication of denominate numbers, we place the multiplier under the lowest denomination of the multiplicand, as

$$\begin{array}{ll} 17 \text{ ft.} & 3 \text{ in.} \\ & 51 \\ \hline 879 \text{ ft.} & 9 \text{ in.} \end{array}$$

and begin at the right to multiply. $51 \times 3 = 153$ in. As there are 12 inches in 1 foot, in 153 in. there are as many feet as 12 is contained times in 153, or 12 feet and 9 inches remaining. Place the 9 inches under the inches, and reserve the 12 feet. 51×17 ft. $= 867$ ft. 867 ft. $+ 12$ ft. (reserved) $= 879$ ft.

879 feet can be reduced to higher denominations by dividing by 3 feet to find the number of yards, and by $5\frac{1}{2}$ yards to find the number of rods.

$$\begin{array}{l} 3\,)\,\underline{8\ 7\ 9} \text{ ft. 9 in.} \\ 5.5\,)\,\underline{2\ 9\ 3} \text{ yd.} \\ \,5\ 3 \text{ rd. } 1\tfrac{1}{2} \text{ yd.} \end{array}$$

Then, answer $= 53$ rd. $1\frac{1}{2}$ yd. 0 ft. 9 in.; or 53 rd. 1 yd. 2 ft. 3 in.

(**109**)

	qt.	pt.	gi.
	3	**1**	3
			4.7
	1 8.2 qt.	0	.1
or,	1 8 qt.	0 pt.	1.7 gi.
or, 4 gal.	2 qt.	0 pt.	1.7 gi. Ans.

Place the multiplier under the lowest denomination of the multiplicand, and proceed to multiply. 4.7×3 gi. $= 14.1$ gi. As 4 gi. $= 1$ pt., there are as many pints in 14.1 gi. as 4 is contained times in $14.1 = 3.5$ pt. and .1 gi. over. Place .1 under gills and carry the 3.5 pt. forward. 4.7×1 pt. $= 4.7$ pt.; $4.7 + 3.5$ pt. $= 8.2$ pt. As 2 pt. $= 1$ qt., there are as many quarts in 8.2 pt. as 2 is contained times in $8.2 = 4.1$ qt. and no pints over. Place a cipher under the pints, and carry the 4.1 qt. to the next product. 4.7×3 qt. $= 14.1$; $14.1 + 4.1 = 18.2$ qt. The answer now is 18.2 qt. 0 pt. .1

gi. Reducing the fractional part of a quart, we have 18 qt. 0 pt. 1.7 gi. (.2 qt. = .2 × 8 = 1.6 gi.; 1.6 + .1 gi. = 1.7 gi.). Then, we can reduce 18 qt. to gallons (18 ÷ 4 = 4 gal. and 2 qt.) = 4 gal. 2 qt. 1.7 gi. Ans.

The answer may be obtained in another and much easier way by reducing all to gills, multiplying by 4.7, and then changing back to quarts and pints. Thus,

$$
\begin{array}{r}
3 \text{ qt.} \\
\times\ 2 \text{ pt.} \\
\hline
6 \text{ pt.} \\
+\ 1 \text{ pt.} \\
\hline
7 \text{ pt.} \\
\times\ 4 \text{ gi.} \\
\hline
28 \text{ gi.} \\
+\ 3 \text{ gi.} \\
\hline
31 \text{ gi.}
\end{array}
$$

3 qt. 1 pt. 3 gi. = 31 gi.
31 gi. × 4.7 = 145.7 gi.

$$
\begin{array}{r}
4\,)\,1\,4\,5.7 \text{ gi.} \\
\hline
2\,)\,3\,6 \quad \text{pt.} + 1.7 \text{ gi.} \\
\hline
1\,8 \quad \text{qt.} + \quad 0 \text{ pt.}
\end{array}
$$

Ans. = 18 qt. 1.7 gi.;
or, 4 gal. 2 qt. 1.7 gi.

(**110**) (3 lb. 10 oz. 13 pwt. 12 gr.) × 1.5 = ?

$$
\begin{array}{r}
3 \text{ lb. } 10 \text{ oz. } 13 \text{ pwt. } 12 \text{ gr.} \\
\times\,1\,2 \\
\hline
3\,6 \quad \text{oz.} \\
+\,1\,0 \\
\hline
4\,6 \quad \text{oz.} \\
\times\quad 2\,0 \\
\hline
9\,2\,0 \text{ pwt.} \\
+\quad 1\,3 \\
\hline
9\,3\,3 \text{ pwt.} \\
\times\quad 2\,4 \\
\hline
2\,2\,3\,9\,2 \text{ gr.} \\
+\quad 1\,2 \\
\hline
2\,2\,4\,0\,4 \text{ gr.}
\end{array}
$$

22.404 gr. × 1.5 = 33,606 gr.

$$
\begin{array}{r}
24\,)\,3\,3\,6\,0\,6 \text{ gr.} \\
\hline
20\,)\,1\,4\,0\,0 \text{ pwt.} + 6 \text{ gr.} \\
\hline
12\,)\,7\,0 \text{ oz.} + 0 \text{ pwt.} \\
\hline
5 \text{ lb.} + 10 \text{ oz.}
\end{array}
$$

Since there are 24 gr. in 1 pwt., in 33,606 gr. there are as many pwt. as 24 is contained times in 33,606, or 1,400 pwt. and 6 gr. remaining. This gives us the number of grains in the answer. We now reduce 1,400 pwt. to higher denominations. Since there are 20 pwt. in 1 oz., in 1,400 pwt. there are as many ounces as 20 is contained times in 1,400, or 70 oz. and 0 pwt. remaining; therefore, there are 0 pwt. in the answer. We reduce 70 oz. to higher denominations. Since there are 12 oz. in 1 lb., in 70 oz. there are as many pounds as 12 is contained times in 70, or 5 lb. and 10 oz. remaining. We can not reduce 5 lb. to any higher denominations. Therefore, our answer is 5 lb. 10 oz. 6 gr.

Another but more complicated way of working this problem is as follows:

lb.	oz.	pwt.	gr.
3	10	13	12
			1.5
4.5	15	19.5	18
or, 4	21	19	30
or, 5	10	0	6 Ans.

To get rid of the decimal in the pounds, reduce .5 of a pound to ounces. Since 1 lb. = 12 oz., .5 of a pound equals .5 lb. × 12 = 6 oz. 6 oz. +.15 oz. = 21 oz. We now have 4 lb. 21 oz. 19.5 pwt. and 18 gr., but we still have a decimal in the column of pwt., so we reduce .5 pwt. to grains to get rid of it. Since 1 pwt. = 24 gr., .5 pwt. = .5 pwt. × 24 = 12 gr. 12 gr. + 18 gr. = 30 gr. We now have 4 lb. 21 oz. 19 pwt. and 30 gr. Since there are 24 gr. in 1 pwt., in 30 gr. there is 1 pwt. and 6 gr. remaining. Place 6 gr. under the column of grains and add 1 pwt. to the pwt. column. Adding 1 pwt., we have 19 + 1 = 20 pwt. Since there are 20 pwt. in 1 oz., we have 1 oz. and 0 pwt. remaining. Write the 0 pwt. under the pwt. column, and reserve the 1 oz. to the oz. column. 21 oz. + 1 oz. = 22 oz. Since there are 12 oz. in 1 lb., in 22 oz. there is 1 lb. and 10 oz. remaining. Write the 10 oz. under the ounce column, and reserve the 1 lb. to add to the lb. column. 4 lb. + 1 lb. (reserved) = 5 lb. Hence, the answer equals 5 lb. 10 oz. 6 gr.

(111) If each barrel of apples contains 2 bu. 3 pk. and 6 qt., then 9 bbl. will contain 9 × (2 bu. 3 pk. 6 qt.).

We write the multiplier under the lowest denomination of the multiplicand, which is quarts in this problem. 9 times 6 qt. equals 54 qt. There are 8 qt. in 1 pk., and in 54 qt. there are as many pecks as 8 is contained times in 54, or 6 pk. and 6 qt. We write the 6 qt. under the column of quarts, and reserve the 6 pk. to add to the product of the pecks. 9 times 3 pk. equals 27 pk.; 27 pk. plus the 6 pk. reserved equals 33 pk. Since there are 4 pk. in 1 bu., in 33 pk. there are as many bushels as 4 is contained times in 33, or 8 bu. and 1 pk. remaining. We write the 1 pk. under the column of pecks, and reserve the 8 bu. for the product of the bushels. 9 times 2 bu. plus the 8 bu. reserved equals 26 bu. Therefore, we find that 9 bbl. contain 26 bu. 1 pk. 6 qt. of apples. Ans.

bu.	pk.	qt.
2	3	6
		9
18	27	5 4
or, 2 6	1	6

(112) (7 T. 15 cwt. 10.5 lb.) × 1.7 = ? When the multiplier is a decimal, instead of multiplying the denominate numbers as in the case when the multiplier is a whole number, it is much easier to reduce the denominate numbers to the lowest denomination given; then, multiply that result by the decimal, and, lastly, reduce the product to higher denominations. Although the correct answer can be obtained by working examples involving decimals in the manner as in the last example, it is much more complicated than this method.

$$
\begin{array}{r}
7 \text{ T. } 15 \text{ cwt. } 10.5 \text{ lb.} \\
\times \quad 2\,0 \\
\hline
1\,4\,0 \quad \text{cwt.} \\
1\,5 \\
\hline
1\,5\,5 \quad \text{cwt.} \\
\times \quad 1\,0\,0 \\
\hline
1\,5\,5\,0\,0 \quad \text{lb.} \\
1\,0.5 \\
\hline
1\,5\,5\,1\,0.5 \text{ lb.}
\end{array}
$$

15,510.5 lb. × 1.7 = 26,367.85 lb.

There are 100 lb. in 1 cwt., and in 26,367.85 lb. there are
as many cwt. as 100 is contained times in 26,367.85, which
equals 263 cwt. and 67.85 lb.

100) 2 6 3 6 7.8 5 lb.   remaining.   Since we have
20) 2 6 3 cwt. + 67.85 lb.   the number of pounds for
        13 T. + 3 cwt.   our answer, we reduce 263
                                          cwt. to higher denominations.
There are 20 cwt. in 1 ton, and in 263 cwt. there are as
many tons as 20 is contained times in 263, or 13 tons and 3
cwt. remaining.   Since we cannot reduce 13 tons any
higher, our answer is 13 T. 3 cwt. 67.85 lb.   Or, since .85
lb. = .85 lb. × 16 = 13.6 oz., the answer may be written
13 T. 3 cwt. 67 lb. 13.6 oz.

(113) 7) 358 A.   57 sq. rd.   6 sq. yd.   2 sq. ft.
             51 A.   31 sq. rd.   0 sq. yd.   8 sq. ft. Ans.

We begin with the highest denomination, and divide each
term in succession by 7.

7 is contained in 358 A. 51 times and 1 A. remaining.
We write the 51 A. under the 358 A. and reduce the remaining 1 A. to square rods = 160 sq. rd.; 160 sq. rd. + the 57
sq. rd. in the dividend = 217 sq. rd.   7 is contained in 217
sq. rd. 31 times and 0 sq. rd. remaining.   7 is not contained
in 6 sq. yd., so we write 0 under the sq. yd. and reduce
6 sq. yd. to square feet.   9 sq. ft. × 6 = 54 sq. ft.   54 sq. ft.
+ 2 sq. ft. in the dividend = 56 sq. ft.   7 is contained in
56 sq. ft. 8 times.   We write 8 under the 2 sq. ft. in the
dividend.

(114) 12) 282 bu.   3 pk.   1 qt.   1 pt.
               23 bu.   2 pk.   2 qt.   ¼ pt. Ans.

12 is contained in 282 bu. 23 times and 6 bu. remaining.
We write 23 bu. under the 282 bu. in the dividend, and
reduce the remaining 6 bu. to pecks = 24 pk. + the 3 pk. in
the dividend = 27 pk.   12 is contained in 27 pk. 2 times and
3 pk. remaining.   We write 2 pk. under the 3 pk. in the
dividend, and reduce the remaining 3 pk. to quarts.   3 pk.
= 24 qt.; 24 qt. + the 1 qt. in the dividend = 25 qt.   12 is
contained in 25 qt. 2 times and 1 qt. remaining.   We write

2 qt. under the 1 qt. in the dividend, and reduce 1 qt. to
pints = 2 pt. + the 1 pt. in the dividend = 3 pt.   3 ÷ 12 =
$\frac{3}{12}$ or $\frac{1}{4}$ pt.

(115) We must first reduce 23 miles to feet before we
can divide by 30 feet. 1 mi. contains 5,280 ft.; hence, 23
mi. contain 5,280 × 23 = 121,440 ft.

121,440 ft. ÷ 30 ft. = 4,048 rails for 1 side of the track.

The number of rails for 2 sides of the track = 2 × 4,048,
or 8,096 rails. Ans.

(116) In this case where both dividend and divisor are
compound, reduce each to the lowest denomination men-
tioned in either and then divide as in simple numbers.

```
      1 bu. 1 pk. 7 qt.              3 5 6 bu. 3 pk. 5 qt.
        × 4                            × 4
       ─────                          ─────
        4 pk.                        1 4 2 4 pk.
       + 1                            + 3
       ─────                          ─────
        5 pk.                        1 4 2 7 pk.
        × 8                            × 8  .
       ─────                          ─────
       4 0 qt.                       1 1 4 1 6 qt.
       + 7                            + 5
       ─────                          ─────
       4 7 qt.                       1 1 4 2 1 qt.
  47 ) 1 1 4 2 1 ( 243
       9 4
       ─────              11,421 qt. ÷ 47 qt. = 243 boxes.
       2 0 2                                       Ans.
       1 8 8
       ─────
       1 4 1
       1 4 1
       ─────
```

(117) We must first reduce 16 square miles to acres.
In 1 sq. mi. there are 640 A., and in 16 sq. mi. there are
16 × 640 A. = 10,240 A.

```
  62 ) 1 0 2 4 0 A.
```
1 6 5 A. 25 sq. rd. 24 sq. yd. 3 sq. ft. 80+sq. in. Ans.

62 is contained in 10,240 A. 165 times and 10 A. remaining. We write 165 A. under the 10,240 A. in the dividend and reduce 10 A. to sq. rd. In 1 A. there are 160 sq. rd., and in 10 A. there are $10 \times 160 = 1,600$ sq. rd. 62 is contained in 1,600 sq. rd. 25 times and 50 sq. rd. remaining. We write 25 sq. rd. in the quotient and reduce 50 sq. rd. to sq. yd. In 1 sq. rd. there are $30\frac{1}{4}$ sq. yd., and in 50 sq. rd. there are 50 times $30\frac{1}{4}$ sq. yd. $= 1,512\frac{1}{2}$ sq. yd. 62 is contained in $1,512\frac{1}{2}$ sq. yd. 24 times and $24\frac{1}{2}$ sq. yd. remaining. In 1 sq. yd. there are 9 sq. ft., and in $24\frac{1}{2}$ sq. yd. there are $24\frac{1}{2} \times 9 = 220\frac{1}{2}$ sq. ft. 62 is contained in $220\frac{1}{2}$ sq.ft. 3 times and $34\frac{1}{2}$ sq. ft. remaining. We write 3 sq. ft. in the quotient and reduce $34\frac{1}{2}$ sq. ft. to sq. in. In 1 sq. ft. there are 144 sq. in., and in $34\frac{1}{2}$ sq. ft. there are $34\frac{1}{2} \times 144 = 4,968$ sq. in. 62 is contained in 4,968 sq. in. 80 times and 8 sq. in. remaining.

We write 80 sq. in. in the quotient.

It should be borne in mind that it is only for the purpose of illustrating the method that this problem is carried out to square inches. It is not customary to reduce any lower than square rods in calculating the area of a farm.

(118) To square a number, we must multiply the number by itself once, that is, use the number twice as a factor. Thus, the second power of 108 is $108 \times 108 = 11,664$. Ans.

$$\begin{array}{r} 108 \\ 108 \\ \hline 864 \\ 108 \\ \hline 11664 \end{array}$$

(119)

```
      1 8 1.2 5
      1 8 1.2 5
    ─────────────
        9 0 6 2 5
      3 6 2 5 0
    1 8 1 2 5
  1 4 5 0 0 0
  1 8 1 2 5
 ─────────────
  3 2 8 5 1.5 6 2 5
      1 8 1.2 5
 ───────────────────
  1 6 4 2 5 7 8 1 2 5
  6 5 7 0 3 1 2 5 0
3 2 8 5 1 5 6 2 5
2 6 2 8 1 2 5 0 0 0
3 2 8 5 1 5 6 2 5
───────────────────────
5 9 5 4 3 4 5.7 0 3 1 2 5
```

The third power of 181.25 equals the number obtained by using 181.25 as a factor three times. Thus, the third power of 181.25 is 181.25 × 181.25 × 181.25 = 5,954,345.703125. Ans.

Since there are 2 decimal places in the multiplier, and 2 in the multiplicand, there are 2 + 2 = 4 decimal places in the first product.

Since there are 4 decimal places in the multiplicand, and 2 in the multiplier, there are 4 + 2 = 6 decimal places in the final product.

(120)

```
          2 7.6 1
          2 7.6 1
       ─────────────
          2 7 6 1
        1 6 5 6 6
      1 9 3 2 7
    5 5 2 2
   ───────────────
      7 6 2.3 1 2 1
          2 7.6 1
   ───────────────────
      7 6 2 3 1 2 1
    4 5 7 3 8 7 2 6
  5 3 3 6 1 8 4 7
1 5 2 4 6 2 4 2
───────────────────────
  2 1 0 4 7.4 3 7 0 8 1
          2 7.6 1
───────────────────────────
  2 1 0 4 7 4 3 7 0 8 1
1 2 6 2 8 4 6 2 2 4 8 6
1 4 7 3 3 2 0 5 9 5 6 7
4 2 0 9 4 8 7 4 1 6 2
───────────────────────────
5 8 1 1 1 9.7 3 7 8 0 6 4 1
```

The fourth power of 27.61 is the number obtained by using 27.61 as a factor four times. Thus, the fourth power of 27.61 is 27.61 × 27.61 × 27.61 × 27.61 = 581,119.73780641. Ans.

Since there are 2 decimal places in the multiplier and 2 in the multiplicand, there are 2 + 2 = 4 decimal places in the first product.

Since there are 4 decimal places in the multiplicand and 2 in the multiplier, there are 4 + 2 = 6 decimal places in the second product.

Since there are 6 decimal places in the multiplicand and 2 in the multiplier, there are 6 + 2 = 8 decimal places in the final product.

(121) (*a*) $106^2 = 106 \times 106 = 11{,}236.$ **Ans.**

$$
\begin{array}{r}
106 \\
106 \\
\hline
636 \\
1060 \\
\hline
11236
\end{array}
$$

(*b*) $\left(182\frac{1}{8}\right)^2 = 182\frac{1}{8} \times 182\frac{1}{8} = 33{,}169.515625.$ **Ans.**

$\frac{1}{8} = 8\,\overline{)1.000}$
$\phantom{\frac{1}{8} = 8)}.125$

$$
\begin{array}{r}
182.125 \\
182.125 \\
\hline
910625 \\
364250 \\
182125 \\
364250 \\
1457000 \\
182125 \\
\hline
33169.515625
\end{array}
$$

Since there are 3 decimal places in the multiplier and 3 in the multiplicand, there are $3 + 3 = 6$ decimal places in the product.

(*c*) $.005^2 = .005 \times .005 = .000025.$ **Ans.**

$$
\begin{array}{r}
.005 \\
.005 \\
\hline
.000025 \quad \text{Ans.}
\end{array}
$$

Since there are 3 decimal places in the multiplicand and 3 in the multiplier, there are $3 + 3 = 6$ decimal places in the product.

(*d*) $.0063^2 = .0063 \times .0063 = .00003969.$ **Ans.**

$$
\begin{array}{r}
.0063 \\
.0063 \\
\hline
189 \\
378 \\
\hline
.00003969 \quad \text{Ans.}
\end{array}
$$

Since there are 4 decimal places in the multiplicand and 4 in the multiplier, there are $4 + 4 = 8$ decimal places in the product.

(*e*) $10.06^2 = 10.06 \times 10.06 = 101.2036.$ **Ans.**

$$
\begin{array}{r}
10.06 \\
10.06 \\
\hline
6036 \\
100600 \\
\hline
101.2036
\end{array}
$$

Since there are 2 decimal places in the multiplicand and 2 in the multiplier, there are $2 + 2 = 4$ decimal places in the product.

(122) (*a*) $753^3 = 753 \times 753 \times 753 = 426,957,777.$ Ans.

```
        753
        753
      2259
     3765
    5271
    567009
       753
   1701027
  2835045
 3969063
 426957777
```

(*b*) $987.4^3 = 987.4 \times 987.4 \times 987.4 = 962,674,279.624.$ Ans.

```
        987.4
        987.4
      39496
     69118
    78992
   88866
   974958.76
        987.4
   389983504
  682471132
 779967008
877462884
962674279.624
```

Since there is 1 decimal place in the multiplicand and 1 in the multiplier, there are $1+1=2$ decimal places in the first product.

Since there are 2 decimal places in the multiplicand and one in the multiplier, there are $2+1=3$ decimal places in the final product.

(*c*) $.005^3 = .005 \times .005 \times .005 = .000000125.$ Ans.

Since there are 3 decimal places in the multiplicand and 3 in the multiplier, there are $3+3=6$ decimal places in the first product; but, as there are only 2 figures in the product, we prefix four ciphers to make the six decimal places.

```
    .005
    .005
    .000025
    .005
    .000000125
```

Since there are six decimal places in the multiplicand and 3 in the multiplier, there are $6+3=9$ decimal places in the final product. In this case we prefix six ciphers to form the nine decimal places.

(d) $.4044^3 = .4044 \times .4044 \times .4044 = .066135317184$ Ans.

```
        .4044
        .4044
      ───────
        16176
        16176
       161760
      ───────
      .16353936
        .4044
      ─────────
      65415744
      65415744
     654157440
    ──────────
    ..066135317184
```

Since there are 4 decimal places in the multiplicand and 4 in the multiplier, there are $4 + 4 = 8$ decimal places in the first product.

Since there are 8 decimal places in the second multiplicand and 4 in the multiplier, there are $8 + 4 = 12$ decimal places in the final product; but, as there are only 11 figures in the product, we prefix 1 cipher to make 12 decimal places.

(123) $2^5 = 2 \times 2 \times 2 \times 2 \times 2 = 32.$ Ans.

(124) $3^4 = 3 \times 3 \times 3 \times 3 = 81.$ Ans.

(125)   (a) $67.85^2 = 67.85 \times 67.85 = 4,603.6225.$ Ans.

```
       67.85
       67.85
      ──────
       33925
       54280
       47495
       40710
      ──────
     4603.6225   Ans.
```

Since there are 2 decimal places in the multiplier and 2 in the multiplicand, there are $2 + 2 = 4$ decimal places in the product.

(b) $967,845^2 = 967,845 \times 967,845 = 936,723,944,025.$ Ans

```
          967845
          967845
        ────────
         4839225
         3871380
         7742760
         6774915
         5807070
         8710605
       ──────────
       936723944025
```

(*c*) A fraction may be raised to any power by raising both numerator and denominator to the required term.

$$\text{Thus, } \left(\frac{3}{8}\right)^2 = \frac{3}{8} \times \frac{3}{8} = \frac{3 \times 3}{8 \times 8} = \frac{9}{64}. \quad \text{Ans.}$$

(*d*) $\left(\frac{1}{4}\right)^2 = \frac{1}{4} \times \frac{1}{4} = \frac{1 \times 1}{4 \times 4} = \frac{1}{16}.$ Ans.

(126) (*a*) $5^{10} = 5 \times 5 \times 5 \times 5 \times 5 \times 5 \times 5 \times 5 \times 5 \times 5 = 9,765,625.$ Ans.

(*b*) $9^5 = 9 \times 9 \times 9 \times 9 \times 9 = 59,049.$ Ans.

5	9
5	9
25	81
5	9
125	729
5	9
625	6561
5	9
3125	59049
5	
15625	
5	
78125	
5	
390625	
5	
1953125	
5	
9765625	

(127) (*a*) $1.2^4 = 1.2 \times 1.2 \times 1.2 \times 1.2 = 2.0736.$ Ans.

Since there is 1 decimal place in the multiplicand and 1 in the multiplier, we must point off $1 + 1 = 2$ decimal places in the first product.

Since there are 2 decimal places in the second multiplicand and 1 in the multiplier, we must point off $2+1=3$
decimal places in the second product.

Since there are 3 decimal places in the third multiplicand
and 1 in the multiplier, we must point off $3+1=4$ decimal
places in the final product.

$$
\begin{array}{r}
1.2 \\
1.2 \\
\hline
2\ 4 \\
1\ 2 \\
\hline
1.4\ 4 \\
1.2 \\
\hline
2\ 8\ 8 \\
1\ 4\ 4 \\
\hline
1.7\ 2\ 8 \\
1.2 \\
\hline
3\ 4\ 5\ 6 \\
1\ 7\ 2\ 8 \\
\hline
2.0\ 7\ 3\ 6
\end{array}
$$

(b) $11^6 = 11 \times 11 \times 11 \times 11 \times 11 \times 11 = 1,771,561.$ Ans

$$
\begin{array}{r}
1\ 1 \\
1\ 1 \\
\hline
1\ 2\ 1 \\
1\ 1 \\
\hline
1\ 3\ 3\ 1 \\
1\ 1 \\
\hline
1\ 4\ 6\ 4\ 1 \\
1\ 1 \\
\hline
1\ 6\ 1\ 0\ 5\ 1 \\
1\ 1 \\
\hline
1\ 7\ 7\ 1\ 5\ 6\ 1
\end{array}
$$

(c) $1' = 1 \times 1 \times 1 \times 1 \times 1 \times 1 \times 1 = 1.$ Ans.

(d) $.01' = .01 \times .01 \times .01 \times .01 = .00000001.$ Ans.

Since there are 2 decimal places in the multiplicand and 2 in the multiplier, we must point off $2 + 2 = 4$ decimal places in the first product; but, as there is only 1 figure in the product, we prefix 3 ciphers to make the 4 necessary decimal places.

```
     .01
     .01
   ------
   .0001
     .01
   --------
 .000001
     .01
 ----------
.00000001
```

Since there are 4 decimal places in the second multiplicand and 2 in the multiplier, we must point off $4 + 2 = 6$ decimal places in the second product. It is necessary to prefix 5 ciphers to make 6 decimal places.

Since there are 6 decimal places in the third multiplicand and 2 in the multiplier, we must point off $6 + 2 = 8$ decimal places in the product. It is necessary to prefix 7 ciphers to make 8 decimal places in the final product.

(e) $.1' = .1 \times .1 \times .1 \times .1 \times .1 = .00001.$ Ans.

Since there is 1 decimal place in the multiplicand and 1 in the multiplier, we must point off $1 + 1 = 2$ decimal places in the first product. It is necessary to prefix 1 cipher to the product.

```
    .1
    .1
  -----
   .01
    .1
  -----
  .001
    .1
  ------
 .0001
    .1
 -------
.00001
```

Since there are 2 decimal places in the second multiplicand and 1 in the multiplier, we must point off $2 + 1 = 3$ decimal places in the second product. It is necessary to prefix 2 ciphers to the second product.

Since there are 3 decimal places in the third multiplicand and 1 in the multiplier, we must point off $3 + 1 = 4$ decimal places in the third product. It is necessary to prefix 3 ciphers to this product.

Since there are 4 decimal places in the fourth multiplicand and 1 in the multiplier, we must point off $4 + 1$ or 5 decimal places in the final product. It is necessary to prefix 4 ciphers to this product.

(128) (*a*) .0133' $=$.0133 $\times$.0133 $\times$ .0133 $=$.000002352637.

Ans.

Since there are 4 decimal places in the multiplicand and 4 in the multiplier, we must point off $4 + 4 = 8$ decimal

.0133
.0133
—
399
399
133
—
.00017689
.0133
—
53067
53067
17689
—
.000002352637

places in the product; but, as there are only 5 figures in the product, we prefix three ciphers to form the eight necessary decimal places in the first product.

Since there are 8 decimal places in the multiplicand and 4 in the multiplier, we must point off $8 + 4 = 12$ decimal places in the product; but, as there are only 7 figures in the product, we prefix 5 ciphers to make the 12 necessary decimal places in the final product.

(*b*)   301.011' $= 301.011 \times 301.011 \times 301.011 =$

27,273,890.942264331. Ans.

301.011
301.011
—
301011
301011
3010110
9030330
—
90607.622121
301.011
—
90607622121
90607622121
90607622120
2718228663630
—
27273890.942264331

Since there are 3 decimal places in the multiplicand and 3 in the multiplier, we must point off $3 + 3 = 6$ decimal places in the first product.

Since there are 6 decimal places in the multiplicand and 3 in the multiplier, we must point off $6 + 3 = 9$ decimal places in the final product.

(c) $\left(\frac{1}{8}\right)^3 = \frac{1}{8} \times \frac{1}{8} \times \frac{1}{8} = \frac{1 \times 1 \times 1}{8 \times 8 \times 8} = \frac{1}{512}.$ Ans.

(d) To find any power of a mixed number, first reduce it to an improper fraction, and then multiply the numerators together for the numerator of the answer, and multiply the denominators together for the denominator of the answer.

$$\left(3\frac{3}{4}\right)^3 = \frac{15}{4} \times \frac{15}{4} \times \frac{15}{4} = \frac{15 \times 15 \times 15}{4 \times 4 \times 4} = \frac{3,375}{64} = 52.734+.$$
Ans.

$$3\frac{3}{4} = \frac{3 \times 4 + 3}{4} = \frac{12 + 3}{4} = \frac{15}{4}.$$

```
      15          64)3375.000(52.734+
      15             320
     ───             ───
      75             175
      15             128
     ───             ───
     225             470
      15             448
    ────             ───
    1125             220
     225             192
    ────             ───
    3375             280
                     256
                     ───
                      24
```

Since *three* ciphers were annexed to the dividend, *three* decimal places must be pointed off in the quotient. It is easy to see that the next figure will be a 3; hence, write the sign +, as shown.

(129) Evolution is the reverse of involution. In involution we find the *power* of a number by multiplying the number by itself one or more times, while in evolution we find the *number* or *root* which was multiplied by itself one or more times to make the power.

(130)   (*a*)

$$
\begin{array}{r}
1 \\
1 \\
\hline
2\,0 \\
8 \\
\hline
2\,8 \\
8 \\
\hline
3\,6\,0 \\
6 \\
\hline
3\,6\,6 \\
6 \\
\hline
3\,7\,2\,0 \\
7 \\
\hline
3\,7\,2\,7 \\
7 \\
\hline
3\,7\,3\,4
\end{array}
$$

$$\sqrt{3'4\,8'6\,7'8\,4.4\,0'1\,0} = 1867.29 \div \quad \text{Ans.}$$

$$
\begin{array}{r}
1 \\
\hline
2\,4\,8 \\
2\,2\,4 \\
\hline
2\,4\,6\,7 \\
2\,1\,9\,6 \\
\hline
2\,7\,1\,8\,4 \\
2\,6\,0\,8\,9 \\
\hline
\end{array}
$$

$$
3734\,)\,1\,0\,9\,5.0\,0\,0\,(.293 \text{ or } .29+
$$

$$
\begin{array}{r}
7\,4\,6\,8 \\
\hline
3\,4\,8\,2\,0 \\
3\,3\,6\,0\,6 \\
\hline
1\,2\,1\,4\,0
\end{array}
$$

EXPLANATION.—Applying the short method described in Art. **272**, we extract the root by the regular method to four figures, since there are six figures in the answer, and $6 \div 2 + 1 = 4$. The last remainder is 1095, and the last trial divisor (with the cipher omitted) is 3734. Dividing 1095 by 3734, as shown, the quotient is .293 +, or .29 + using two figures. Annexing to the root, gives 1,867.29 + . Ans.

$(\mathring{b})$ (a) 3 $\sqrt{9'0\,0'0\,0'9\,9.4\,0'0\,9'0\,0} = 3000.0165 +$ Ans.

3 (b) 9

(d) 6 0 (c) 0 0 0 0 9 9 4 0 0 9

     0   6 0 0 0 0 1

    6 0 0   3 9 4 0 0 8 0 0

     0   3 6 0 0 0 1 5 6

   6 0 0 0   3 4 0 0 6 4 4

     0

  6 0 0 0 0

     0

6 0 0 0 0 0

      1

6 0 0 0 0 1

      1

6 0 0 0 0 2 0

       6

6 0 0 0 0 2 6

EXPLANATION.—Beginning at the decimal point we point off the whole number into periods of *two* figures each, proceeding from *right* to *left;* also, point off the decimal into periods of *two* figures each, proceeding from *left* to *right.* The largest number whose square is contained in the first period, 9, is 3; hence, 3 is the first figure of the root. Place 3 at the left, as shown at (a), and multiply it by the first figure in the root, or 3. The result is 9. Write 9 under the first period, 9, as at (b), subtract, and there is no remainder. Bring down the next period, which is 00, as shown at (c). Add the root already found to the 3 at (a), obtaining 6, and annex a cipher to this 6, thus making it 60, which is the *trial divisor*, as shown at (d). Divide the dividend (c) by the trial divisor, and obtain 0 as the next figure in the root. Write 0 in the root, as shown, and also add it to the trial divisor, 60, and annex a cipher, thereby making the next trial divisor 600. Bring down the next period, 00, annex it to the dividend already obtained, and divide it by the trial divisor. 600 is contained in 0000, 0 times, so we place another cipher in the root. Write 0 in the root, as shown, and also add it to the trial divisor, 600, and annex a cipher, thereby making the next trial divisor 6,000. Bring down the next period, 99. The trial divisor 6,000 is contained in 000099, 0 times, so we place 0 as the next figure in the root, as shown, and also add it to the trial divisor 6,000, and annex a cipher, thereby making the next trial divisor 60,000. Bring down the next period, 40, and annex it to the dividend already obtained to form the new dividend, 00009940, and divide it by the trial divisor 60,000. 60,000 is contained in 00009940, 0 times, so we place another cipher in the root, as shown, and also add it to the trial divisor 60,000, and annex one cipher, thereby making the next trial divisor 600,000. Bring down the next period, 09, and annex it to the dividend already obtained to form the new dividend, 0000994009, and divide it by the trial divisor 600,000. 600,000 is contained in 0000994009 once, so we place 1 as the next figure in the root, and also add it to the trial divisor 600,000, thereby making the complete divisor 600,001. Multiply the complete divisor, 600,001, by 1, the sixth figure in the root, and subtract the result obtained from the dividend. The remainder is 394,008, to which we annex the next period, 00, to form the next new dividend, or 39,400,800. Add the sixth figure of the root, or 1, to the divisor 600,001, and annex a cipher, thus obtaining 6,000,020 as the next trial divisor. Dividing 39,400,800 by 6,000,020, we find 6 to be the next figure of the root. Adding this last figure, 6, to the trial divisor, we obtain 6,000,026 for our next complete divisor, which, multiplied by the last figure of the root, or 6, gives 36,000,156, which write under 39,400,800 and subtract. Since there is a remainder, it is clearly evident that the given power is not a perfect square, so we place + after the root. Since the next figure is 5, the answer is 3,000.017 —.

In this problem there are *seven* periods—four in the whole number and three in the decimal—hence, there will be *seven* figures in the root, *four* figures constituting the whole number, and three figures the decimal of the root. Hence,

$$\sqrt{9,000,099.4009} = 3,000.017 -.$$

(c)
$$\sqrt{.00'12'25} = .035. \quad \text{Ans.}$$

3	0 0
3	
6 0	1 2
5	9
6 5	3 2 5
	3 2 5

Pointing off periods, we find that the first period is composed of ciphers; hence, the first figure of the root will be a cipher. No further explanation is necessary, since this problem is solved in a manner exactly similar to the problem solved in Art. **264.** Since there are *three* decimal periods in the power, there will be three decimal figures in the root.

(131) (*a*)
$$\sqrt{1'07'95.21} = 103.9 \quad \text{Ans.}$$

1	1
1	0 7 9 5
2 0	6 0 9
0	1 8 6 21
2 0 0	1 8 6 21
3	
2 0 3	
3	
2 0 6 0	
9	
2 0 6 9	

(*b*)
$$\sqrt{7'30'08.04} = 270.2 \quad \text{Ans}$$

2	4
2	3 3 0
4 0	3 2 9
7	1 0 8 0 4
4 7	1 0 8 0 4
7	
5 4 0 0	
2	
5 4 0 2	

(c)

```
        9
        9
       ───
      180
        4
      ───
      184
        4
      ────
     1880
        8
      ────
     1888
        8
      ────
     1896
```

$\sqrt{90.00'00'00} = 9.487 -$ **Ans.**

```
       81
      ────
      900
      736
     ─────
     16400
     15104
```

1896) 1 2 9 6.0 0 (.68 + or .7 --

```
          11376
         ──────
          15840
          15168
         ──────
```

Having found the first three figures, we find the fourth by division, as shown.

(d) $\sqrt{.09} = .3.$ **Ans.**

(132)   (a)

```
      6          3 6
      6          7 2
     ──        ─────
     12         10800
      6          1504
    ───        ──────
    180         12304
      8          1568
    ───        ──────
    188        1387200
      8          18441
    ───        ───────
    196        1405641
      8          18522
   ────        ───────
   2040        1424163
      9
   ────
   2049
      9
   ────
   2058
      9
   ────
   2067
```

$\sqrt{.32'7'68'0'000} = .6894 +$ **Ans.**

```
       216
      ──────
      111680
       98432
     ────────
     13248000
     12650769
    ─────────
```

1424163) 5 9 7 2 3 1.0 0 (.41 + or .4 +

```
             5696652
            ────────
             2756580
             1424163
            ────────
```

Here we find the first three figures in the regular way, and the fourth figure by the short method. See Art. **284.**

EXPLANATION.—(1) When extracting the *cube* root we divide the power into periods of three figures each. Always begin at the decimal point, and proceed to the *left* in pointing off the whole number, and to the *right* in pointing off the decimal. In this power $\sqrt[3]{.32768}$, a cipher must be annexed to 68 to complete the second decimal period. Cipher periods may now be annexed until the root has as many figures as desired.

(2) We find by trial that the largest number whose cube is contained in the first period, 327, is 6. Write 6 as the first figure of the root, also at the extreme left at the head of column (1). Multiply the 6 in column (1) by the first figure of the root, 6, and write the product 36 at the head of column (2). Multiply the number in column (2) by the first figure of the root, 6, and write the product 216 under the figures in the first period. Subtract and bring down the next period 680; annex it to the remainder 111, thereby obtaining 111,680 for a new dividend. Add the first figure of the root, 6, to the number in column (1), obtaining 12, which we call the *first correction;* multiply the first correction 12 by the first figure of the root, and we obtain 72 as the product, which, added to 36 of column (2), gives 108. Annexing two ciphers to 108, we have 10,800 for the trial divisor. Dividing the dividend by the trial divisor, we see that it is contained about 8 times, so we write 8 as the second figure of the root. Add the first figure of the root to the first correction, and we obtain 18 as the *second correction.* To this annex *one* cipher, and add the second figure of the root, and we obtain 188. This, multiplied by the second figure of the root, 8, equals 1,504, which, added to the trial divisor 10,800, forms the *complete divisor* 12,304. Multiplying the complete divisor 12,304 by 8, the second figure of the root, the result is 98,432. Write 98,432 under the dividend 111,680; subtract, and there is a remainder of 13,248. To this remainder annex the next period 000, thereby obtaining 13,248,000 for the next new dividend.

(3) Adding the second figure of the root, 8, to the number in column (1), 188, we have 196 for the *first new*

correction. This, multiplied by the second figure of the root, 8, gives 1,568. Adding this product to the last complete divisor, and annexing two ciphers, gives 1,387,200 for the next trial divisor. Adding the second figure of the root, 8, to the first new correction, 196, we obtain 204 for the *new second correction.* Dividing the dividend by the trial divisor 1,387,200, we see that it is contained about 9 times. Write 9 as the third figure of the root. Annex *one* cipher to the *new second correction,* and to this add the third figure of the root, 9, thereby obtaining 2,049. This, multiplied by 9, the third figure of the root, equals 18,441, which, added to the trial divisor, 1,387,200, forms the complete divisor 1,405,641. Multiplying the complete divisor by the third figure of the root, 9, and subtracting, we have a remainder of 597,231. We then find the fourth figure by division, as shown.

(b)	4		1 6		$\sqrt{7\,4'0\,8\,8} = 42$ Ans.
	4		3 2		6 4
	8		4 8 0 0		1 0 0 8 8
	4		2 4 4		1 0 0 8 8
	1 2 0		5 0 4 4		
	2				
	1 2 2				

(c)	4		1 6		$\sqrt[3]{9\,2'4\,1\,6} = 45.212 -$ Ans.
	4		3 2		6 4
	8		4 8 0 0		2 8 4 1 6
	4		6 2 5		2 7 1 2 5
	1 2 0		5 4 2 5		1 2 9 1 0 0 0
	5		6 5 0		1 2 2 0 4 0 8
	1 2 5		6 0 7 5 0 0		612912)70592.000(.115
	5		2 7 0 4		6 1 2 9 1 2
	1 3 0		6 1 0 2 0 4		9 3 0 0 8 0
	5		2 7 0 8		6 1 2 9 1 2
	1 3 5 0		6 1 2 9 1 2		3 1 7 1 6 8 0
	2				3 0 6 4 5 6 0
	1 3 5 2				1 0 7 1 2 0
	2				
	1 3 5 4				

(d)	7	49	$\sqrt[3]{.373248} = .72$ Ans.

$$\sqrt[3]{.373248} = .72 \quad \text{Ans.}$$

7	49
7	98
‾14	‾14700
7	424
‾210	‾15124
2	
‾212	

343
‾30248
30248
‾

(133)

$$\sqrt[3]{2.000000000} = 1.259921 + \quad \text{Ans}$$

1	1	1
1	2	‾1000
‾2	‾300	728
1	64	‾272000
‾30	‾364	225125
2	68	‾46875000
‾32	‾43200	42491979
2	1825	‾
‾34	‾45025	4755243) 4383021.000 (9217 or .922−
2	1850	42797187
‾360	‾4687500	‾10330230
5	33831	9510486
‾365	‾4721331	‾8197440
5	33912	4755243
‾370	‾4755243	‾34421970
5		
‾3750		
9		
‾3750		
9		
‾3768		

This example shows what a great saving of figures is effected by using the short method. The figures obtained by the division are 9217, thus making the last figures of the answer 922, according to Art. **272.** This is not correct in this case; the true answer to eight decimal places being 1.25992104 +; hence, the first three figures found by division should be used in this case. The reason for the apparent failure of the method in this case to give the seventh figure of the root correctly is because the fifth figure (the first obtained by division) is 9. Whenever the first figure obtained by division is 8 or 9, it is better to carry the root process one place further, before applying Art. **272,** if it is desired to obtain absolutely correct results.

(134) (a)

```
1          1              ∛1'7 5 8.4 1 6'7 4 3 = 12.07   Ans.
1          2                 1
─          ───              ────
2          3 0 0             7 5 8
1            6 4             7 2 8
──         ─────            ──────────
3 0        3 6 4            3 0 4 1 6 7 4 3
   2          6 8           3 0 4 1 6 7 4 3
──         ───────         ─────────────
3 2        4 3 2 0 0 0 0
   2          2 5 2 4 9
──         ─────────
3 4        4 3 4 5 2 4 9
   2
────
3 6 0 0
      7
────
3 6 0 7
```

(b)
```
1          1              ∛1'1 9 1'0 1 6 = 106   Ans.
   1          2              1
──         ─────            ────────
2          3 0 0 0 0        1 9 1 0 1 6
   1          1 8 3 6       1 9 1 0 1 6
────       ─────────       ───────────
3 0 0      3 1 8 3 6
      6
────
3 0 6
```

(c) $\sqrt[3]{\dfrac{4}{32}} = \sqrt[3]{\dfrac{1}{8}} = \dfrac{\sqrt[3]{1}}{\sqrt[3]{8}} = \dfrac{1}{2}.$ Ans.

(d) $\sqrt[3]{\dfrac{27}{512}} = \dfrac{\sqrt[3]{27}}{\sqrt[3]{512}} = \dfrac{3}{8}.$ Ans.

(135)

$$\sqrt[3]{3.000000000} = 1.442250 - \quad \text{Ans.}$$

```
    1        1                 1
    1        2            ─────────
  ───        ───            2 0 0 0
    2        3 0 0          1 7 4 4
    1        1 3 6        ───────────
  ────       ─────          2 5 6 0 0 0
   3 0       4 3 6          2 4 1 9 8 4
    4        1 5 2        ─────────────
  ────       ─────          1 4 0 1 6 0 0 0
   3 4       5 8 8 0 0      1 2 4 5 8 8 8 8
    4        1 6 9 6    ─────────────────────
  ────       ─────────  6238092) 1 5 5 7 1 1 2.0 0 0 ( .2496 or .250 −
   3 8       6 0 4 9 6            1 2 4 7 6 1 8 4
    4        1 7 1 2            ─────────────────
  ─────      ─────────            3 0 9 4 9 3 6 0
   4 2 0     6 2 2 0 8 0 0        2 4 9 5 2 3 6 8
    4            8 6 4 4        ─────────────────
  ─────      ─────────────        5 9 9 6 9 9 2 0
   4 2 4     6 2 2 9 4 4 4        5 6 1 4 2 8 2 8
    4            8 6 4 8        ─────────────────
  ─────      ─────────────          3 8 2 7 0 9 2
   4 2 8     6 2 3 8 0 9 2
    4
  ─────
   4 3 2 0
    2
  ─────
   4 8 2 2
    2
  ─────
   4 3 2 4
```

(136) (a)

```
  1      √1'23.21 = 11.1 Ans.
  1        1
 ───      ───
  2 0     2 3
  1       2 1
 ───      ─────
  2 1     2 2 1
  1       2 2 1
 ─────
  2 2 0
  1
 ─────
  2 2 1
```

(b)

```
  1      √1'14.9 2'10 = 10.72 +
  1        1                       Ans.
 ─────    ─────────
  2 0 0   1 4 9 2
    7     1 4 4 9
 ─────    ─────────
  2 0 7     4 3 1 0
    7       4 2 8 4
 ─────    ─────────
  2 1 4 0       2 6
    2
 ─────
  2 1 4 2
```

(c)

```
  7      √50'26'81 = 709 Ans.
  7        4 9
 ─────    ─────────
  1 4 0   1 2 6 8 1
    0     1 2 6 8 1
 ───────
  1 4 0 0
    9
 ───────
  1 4 0 9
```

(d)

```
  2      √.00'04'12'09 = .0203
  2        0 0                  Ans.
 ─────    ─────
  4 0 0   0 4
    3       4
 ─────    ─────────
  4 0 3   1 2 0 9
          1 2 0 9
```

(137) *(a)*

1	1	$\sqrt[3]{.006'500'000} = .18663 -$ Ans
1	2	1
—	—	—
2	300	5500
1	304	4832
—	—	—
30	604	668000
8	368	602856
—	—	—
38	97200	103788) 65144.00 (.627 or .63 —
8	3276	622728
—	—	—
46	100476	287120
8	3312	207576
—	—	—
540	103788	79544
6		
—		
546		
6		
—		
552		

(b)

2	4	$\sqrt[3]{.021'000'000} = .2759-$ Ans.
2	8	8
—	—	—
4	1200	13000
2	469	11683
—	—	—
60	1669	1317000
7	518	1113875
—	—	—
67	218700	226875) 203125.0 (.89 or .9—
7	4075	1815000
—	—	—
74	222775	216250
7	4100	
—	—	
810	226875	
5		
—		
815		
5		
—		
820		

(c)

2	4	$\sqrt[3]{8'0\,3\,6'0\,5\,4'0\,2\,7}$ = 2,003 Ans.
2	8	8
4	12000000	036054027
2	18009	36054027
6000	12018009	
3		
6003		

(d)

1	1	$\sqrt[3]{.000'004'096}$ = .016 Ans.
1	2	000
2	300	004
1	216	1
30	516	3096
6		3096
36		

(e)

2	4	$\sqrt[3]{17.000'000}$ = 2.5713— Ans.
2	8	8
4	1200	9000
2	325	7625
60	1525	1375000
5	350	1349593
65	187500	198147) 25407.00 (.128 or .13—
5	5299	198147
70	192799	559230
5	5348	396294
750	198147	162936
7		
757		
7		
764		

(138) (*a*) In this example the index is 4, and equals 2×2. The root indicated is the fourth root, hence the square root must be extracted twice. Thus, $\sqrt[4]{} = \sqrt{}$ of the $\sqrt{}$ and $\sqrt[4]{6561} = \sqrt{\sqrt{6561}} = \sqrt{81} = 9$. Ans.

$$
\begin{array}{ccc}
8 & \sqrt{6\,5'6\,1} = 81 & \sqrt{81} = 9 \quad \text{Ans.} \\
8 & 6\,4 & \\
\hline
1\,6\,0 & 1\,6\,1 & \\
1 & 1\,6\,1 & \\
\hline
1\,6\,1 & &
\end{array}
$$

(*b*) In this example the index is 6, and 6 equals 3×2 or 2×3. The root indicated is the sixth root; hence, extract both the square and cube root, it making no particular difference as to which root is extracted first. Thus,

$$\sqrt[6]{} = \sqrt[3]{} \text{ of the } \sqrt{}, \text{ or } \sqrt{} \text{ of the } \sqrt[3]{}.$$

Hence, $\sqrt[6]{117,649} = \sqrt[3]{\sqrt{117,649}} = \sqrt[3]{343} = 7$. Ans.

$$
\begin{array}{ccc}
3 & \sqrt{11'7\,6'4\,9} = 3\,4\,3 & \sqrt[3]{3\,4\,3} = 7 \text{ Ans.} \\
3 & 9 & \\
\hline
6\,0 & 2\,7\,6 & \\
4 & 2\,5\,6 & \\
\hline
6\,4 & 2\,0\,4\,9 & \\
1 & 2\,0\,4\,9 & \\
\hline
6\,8\,0 & & \\
3 & & \\
\hline
6\,8\,3 & &
\end{array}
$$

(*c*) $\sqrt[6]{.000064} = \sqrt[3]{\sqrt{.000064}} = .2$. Ans.

$\sqrt{.000064} = .008$. $\sqrt[3]{.008} = .2$. Hence, $\sqrt[6]{.000064} = .2$.
Ans.

(d) $\sqrt{\dfrac{3}{8}} = ?$ $\dfrac{3}{8} = .375$, since $8)\overline{3.000}$

$\qquad\qquad\qquad\qquad\qquad\qquad\qquad .375$

7	4 9	$\sqrt{.375'000'000} = .72112+$ Ans.
7	9 8	3 4 3
1 4	1 4 7 0 0	3 2 0 0 0
7	4 2 4	3 0 2 4 8
2 1 0	1 5 1 2 4	1 7 5 2 0 0 0
2	4 2 8	1 5 5 7 3 6 1
2 1 2	1 5 5 5 2 0 0	1559523) 1 9 4 6 3 9.0 0 (.124 or .12 +
2	2 1 6 1	1 5 5 9 5 2 3
2 1 4	1 5 5 7 3 6 1	3 8 6 8 6 7 0
2	2 1 6 2	3 1 1 9 0 4 6
2 1 6 0	1 5 5 9 5 2 3	7 4 9 6 2 4
1		
2 1 6 1		
1		
2 1 6 2		

Hence, $\sqrt{\dfrac{3}{8}} = .72112 +$. Ans.

(139) (a) $\sqrt{\dfrac{1225}{5476}} = \dfrac{\sqrt{1225}}{\sqrt{5476}}$.

3	$\sqrt{12'25} = 35$
3	9
6 0	3 2 5
5	3 2 5
6 5	

Hence, $\sqrt{\dfrac{1225}{5476}} = \dfrac{35}{74}$. Ans.

7	$\sqrt{54'76} = 74$
7	4 9
1 4 0	5 7 6
4	5 7 6
1 4 4	

(b)

$$\sqrt{.33\,'64} = .58 \text{ Ans.}$$

```
  5
  5
─────
100
  8
─────
108
```

```
 25
864
864
───
```

(c)

$$\sqrt{.10\,'00\,'00\,'00} = .31623- \text{ Ans.}$$

```
  3
  3
────
60
  1
────
61
  1
────
620
  6
────
626
  6
────
632
```

```
  9
100
 61
─────
3900
3756
```

$632\,)\,144.00\,(.227 \text{ or } .23-$

```
1264
────
1760
1264
────
 496
```

(d) $25.0\dfrac{3}{4} = 25.075.$

```
       5
       5
──────────
   10000
       7
──────────
   10007
       7
──────────
  100140
       4
──────────
  100144
       4
──────────
 1001480
       9
──────────
 1001489
```

$$\sqrt{25.07\,'50\,'00\,'00\,'00} = 5.00749+ \text{ Ans.}$$

```
    25
──────────
   075000
    70049
──────────
   495100
   400576
──────────
  9452400
  9013401
──────────
   438999
```

(e) $.000\dfrac{4}{9} = .0004444444+.$

```
      2
      2
──────────
     40
      1
──────────
     41
      1
──────────
   4200
      8
──────────
   4208
```

$$\sqrt{.00\,'00\,'44\,'44\,'44\,'44} = .02108+ \text{ Ans}$$

```
    00
──────────
    04
     4
──────────
    44
    41
──────────
   34444
   33664
──────────
     780
```

(140) (a) $\sqrt[4]{2} = \sqrt{\sqrt{2}}$.

```
  1                    √2.0 0'00'0 0'0 0 = 1.41421356 +
  1                        1
 20                      1 0 0
  4                        9 6
 24                          4 0 0
  4                          2 8 1
280                        1 1 9 0 0
  1                        1 1 2 9 6
281                            6 0 4 0 0
  1                            5 6 5 6 4
2820              28284 ) 3 8 3 6.0 0 0 0 ( .13562 or .1356 +
   4                          2 8 2 8 4
2824                        1 0 0 7 6 0
   4                          8 4 8 5 2
28280                       1 5 9 0 8 0
    2                       1 4 1 4 2 0
28282                         1 7 6 6 0 0
    2                         1 6 9 7 0 4
28284                             6 8 9 6

  1                    √1.4 1'4 2'1 3'5 6 = 1.1892 +   Ans.
  1                        1
 20                         4 1
  1                         2 1
 21                         2 0 4 2
  1                         1 8 2 4
220                          2 1 8 1 3
  8                          2 1 3 2 1
228                            4 9 2 5 6
  8                            4 7 5 6 4
2360                           1 6 9 2
  9
2369
  9
23780
   2
23782
```

It is required in this problem to extract the fourth root of 2 to four decimal places; hence, we must extract the square root twice, since $\sqrt[4]{\ } = \sqrt{\ }$ of the $\sqrt{\ }$. In the first operation we carry the root to 8 decimal places, in order to carry the root in the second operation to 4 decimal places.

(b) $\sqrt[6]{6} = \sqrt[3]{\sqrt{6}}.$

2	$\sqrt{6.00'00'00'00'00'00} = 2.4494897428 +$
2	4
40	200
4	176
44	2400
4	1936
480	46400
4	44001
484	239900
4	195936
4880	4396400
9	3919104
4889	489896) 477296.00000 (.974280 or .97428 +
9	4409064
48980	3638960
4	3429272
48984	2096880
4	1959584
489880	1372960
8	979792
489888	3931680
8	3919168
489896	12512

It is required in this problem to find the sixth root of 6; hence it is necessary to extract both the square and cube roots in succession, since the index, 6, equals 2×3 or 3×2. It makes no particular difference as to which root we extract first, but it will be more convenient to extract the square root first. The result has been carried to 10 decimal places; since the answer requires but 5 decimal places, the remaining decimals will not affect the cube root in the fifth decimal place, as the student can see for himself if he will continue the operation.

```
  1        1        √2.449'489'742'800 = 1.34801 —
  1        2          1                         Ans.
 ───      ───        ────
  2       300        1449
  1        99        1197
 ───      ───        ──────
 30       399         252489
  3       108         209104
 ───      ─────       ────────
 33       50700        43385742
  3        1576        43352192
 ───      ─────       ──────────
 36       52276      5451312 ) 33550.000 ( .006 or .01 —
  3        1592                32707872
 ───      ───────              ─────────
390       5386800               842128
  4         32224
 ────     ───────
394       5410024
  4         32288
 ────     ───────
398       5451312
  4
 ────
4020
   8
 ────
4028
   8
 ────
4036
```

(141) *(a)*
```
         1          √3.14'16 = 1.7725 —   Ans.
         1            1
        ───          ───
         20          2 14
          7          1 89
        ───          ─────
         27           2516
          7           2429
        ───          ──────
        340         354 ) 87.00 ( .245 + or  25 —
          7                708
        ───               ─────
        347                1620
          7                1416
        ───               ─────
        354                 204
```

(b)   8
               8
             ─────
             1 6 0
               8
             ─────
             1 6 8
               8
             ─────
             1 7 6 0
                 6
             ─────
             1 7 6 6
                '6
             ─────
             1 7 7 2

$\sqrt{.78'54'00} = .8862 +$ Ans.
   6 4
 ─────
 1 4 5 4
 1 3 4 4
 ─────
   1 1 0 0 0
   1 0 5 9 6
 ─────
 1772) 4 0 4.0 (.22 or .2 +
         3 5 4 4
       ─────
           4 9 6

(142) (a)

1   1
1   2
──   ──
2   3 0 0
1   1 3 6
──   ─────
3 0   4 3 6
  4   1 5 2
──   ─────
3 4   5 8 8 0 0
  4   2 5 5 6
──   ─────
3 8   6 1 3 5 6
  4   2 5 9 2
──   ─────
4 2 0   6 3 9 4 8 0 0
    6   1 7 5 3 6
──   ─────
4 2 6   6 4 1 2 3 3 6
    6   1 7 5 5 2
──   ─────
4 3 2   6 4 2 9 8 8 8
    6
──
4 3 8 0
      4
──
4 3 8 4
      4
──
4 3 8 8

$\sqrt[3]{3.141'600'000} = 1.4646 -$
   1   Ans.
 2 1 4 1
. 1 7 4 4
 ─────
   3 9 7 6 0 0
   3 6 8 1 3 6
 ─────
   2 9 4 6 4 0 0 0
   2 5 6 4 9 3 4 4
 ─────
6429888) 3 8 1 4 6 5 6.0 (.59 or .6−
          3 2 1 4 9 4 4 0
        ─────
          5 9 9 8 1 2 0

(*b*)

8	64	$\sqrt[3]{.523'600'000} = .80599+$ or $.8060-$
8	128	512 Ans.

16	1920000	11600000
8	12025	· 9660125

2400	1932025	1944075)1939875.00 (.99
5	12050	17496675

2405	1944075	1902075.
5		

| 2410 | | · |

(**143**) 11.7 : 13 :: 20 : x.   The product of the means
          11.7 x = 13 × 20   equals the product of the
          11.7 x = 260   extremes.

$$x = \frac{260}{11.7}$$

11.7) 260.000 (22.22+   **Ans.**
       234
       ———
        260
        234
        ———
         260
         234
         ———
          260
          234
          ———
           26

(**144**) (*a*) 20 + 7 : 10 + 8 :: 3 : x.
          27 : 18 :: 3 : x
          27 x = 18 × 3
          27 x = 54
          $x = \dfrac{54}{27} = 2.$ Ans.

(*b*) $12^2 : 100^2 :: 4 : x.$
      144 : 10,000 :: 4 : x
      144 x = 10,000 × 4
      144 x = 40,000

$$x = \frac{40,000}{144}$$

144) 4 0 0 0 0.0 (277.7 + **Ans.**

2 8 8
‾‾‾‾‾
1 1 2 0
1 0 0 8
‾‾‾‾‾
1 1 2 0
1 0 0 8
‾‾‾‾‾
1 1 2 0
1 0 0 8
‾‾‾‾‾
1 1 2

(145) (*a*) $\frac{4}{x} = \frac{7}{21}$, is equivalent to 4 : x :: 7 : 21. The product of the means equals the product of the extremes. Hence,

$$7x = 4 \times 21$$
$$7x = 84$$
$$x = \frac{84}{7} \text{ or } 12. \quad \text{Ans.}$$

(*b*) In like manner,

$\frac{x}{24} = \frac{8}{16}$ is equivalent to x : 24 :: 8 : 16.

$$16x = 24 \times 8$$
$$16x = 192$$
$$x = \frac{192}{16} = 12. \quad \text{Ans.}$$

(*c*) $\frac{2}{10} = \frac{x}{100}$ is equivalent to 2 : 10 :: x : 100.

$$10x = 2 \times 100$$
$$10x = 200$$
$$x = \frac{200}{10} = 20. \quad \text{Ans.}$$

(*d*) $\frac{15}{45} = \frac{60}{x}$ is equivalent to

15 : 45 :: 60 : x.

$$15x = 45 \times 60$$
$$15x = 2,700$$
$$x = \frac{2,700}{15} = 180.$$
Ans.

(*e*) $\frac{10}{150} = \frac{x}{600}$ is equivalent to

10 : 150 :: x : 600.

$$150x = 10 \times 600$$
$$150x = 6,000$$
$$x = \frac{6,000}{150} = 40. \quad \text{Ans.}$$

(146) $x : 5 :: 27 : 12.5.$

$$12.5\,)\,1\,3\,5.0\,(\,10\tfrac{4}{5}\ \text{Ans.}$$
$$\underline{1\,2\,5}$$
$$\underline{1\,0\,0} = \frac{4}{5}$$
$$1\,2\,5$$

(147) $45 : 60 :: x : 24$

$60\,x = 45 \times 24$

$60\,x = 1,080$

$x = \dfrac{1,080}{60} = 18.$ Ans.

(148) $x : 35 :: 4 : 7.$

$7\,x = 35 \times 4$

$7\,x = 140$

$x = \dfrac{140}{7} = 20.$ Ans.

(149) $9 : x :: 6 : 24.$

$6\,x = 9 \times 24$

$6\,x = 216$

$x = \dfrac{216}{6} = 36.$ Ans.

(150)

$\sqrt[3]{1,000} : \sqrt[3]{1,331} :: 27 : x.$

$10 : 11 :: 27 : x.$

$10\,x = 297.$

$x = \dfrac{297}{10} = 29.7.$

Ans.

$\sqrt[3]{1,000} = 10.$

$\sqrt[3]{1,331} = 11.$

$$\begin{array}{r} 1 \\ 1 \\ \hline 2 \\ 1 \\ \hline 30 \\ 1 \\ \hline 31 \end{array}$$

$$\begin{array}{r} 1 \\ 2 \\ \hline 300 \\ 31 \\ \hline 331 \end{array}$$

$$1'3\,3\,1\,(\,11$$
$$\underline{1}$$
$$\begin{array}{r} 331 \\ 331 \\ \hline \end{array}$$

(151) $64 : 81 = 21^2 : x^2.$

Extracting the square root of each term of any proportion does not change its value, so we find that $\sqrt{64} : \sqrt{81} = \sqrt{21^2} : \sqrt{x^2}$ is the same as

$$8 : 9 = 21 : x$$
$$8\,x = 189$$
$$x = 23.625.\ \ \text{Ans.}$$

(152) $7 + 8 : 7 = 30 : x$ is equivalent to

$$15 : 7 = 30 : x.$$
$$15\,x = 7 \times 30$$
$$15\,x = 210$$
$$x = \frac{210}{15} = 14.\ \ \text{Ans.}$$

(153) 2 ft. 5 in. = 29 in.; 2 ft. 7 in. = 31 in. Stating as a direct proportion, $29 : 31 = 2,480 : x$. Now, it is easy to see that x will be greater than 2,480. But x should be less than 2,480, since, when a man lengthens his steps, the number of steps required for the same distance is less; hence, the proportion is an inverse one, and

$$29 : 31 = x : 2,480,$$
$$\text{or,} \quad 31\,x = 71,920;$$
$$\text{whence,} \quad x = 71,920 \div 31 = 2,320 \text{ steps.} \quad \text{Ans.}$$

(154) This is evidently a direct proportion. 1 hr. 36 min. = 96 min.; 15 hr. = 900 min. Hence,

$$96 : 900 = 12 : x,$$
$$\text{or,} \quad 96\,x = 10,800;$$
$$\text{whence,} \quad x = 10,800 \div 96 = 112.5 \text{ mi.} \quad \text{Ans.}$$

(155) This is also a direct proportion; hence,

$$27.63 : 29.4 = .76 : x,$$
$$\text{or,} \quad 27.63\,x = 29.4 \times .76 = 22.344;$$
$$\text{whence,} \quad x = 22.344 \div 27.63 = .808 + \text{lb.} \quad \text{Ans.}$$

(156) 2 gal. 3 qt. 1 pt. = 23 pt.; 5 gal. 3 qt. = 46 pt. Hence,

$$23 : 46 = 5 : x,$$
$$\text{or,} \quad 23\,x = 46 \times 5 = 230;$$
$$\text{whence,} \quad x = 230 \div 23 = 10 \text{ days.} \quad \text{Ans.}$$

(157) Stating as a direct proportion, and squaring the distances, as directed by the statement of the example, $6^2 : 12^2 = 24 : x$. Inverting the second couplet, since this is an inverse proportion,

$$6^2 : 12^2 = x : 24.$$

Dividing both terms of the first couplet (see Art. **310**) by 6

$$1^2 : 2^2 = x : 24; \text{ or } 1 : 4 = x : 24;$$
$$\text{whence,} \quad 4\,x = 24, \text{ or } x = 6 \text{ degrees.} \quad \text{Ans.}$$

(158) Taking the dimensions as the causes,

$$\begin{array}{c|c} 12 & 15 \\ 4 & 5 \\ 2 & 2 \\ 3 & 6 \end{array} = 12 \;\middle|\; x, \text{ whence, } 2\,x = 75, \text{ or, } x = \$37.50.$$
Ans.

(159) 2 hr. = 120 min.; 14 hr. 28 min. = 868 min.

Hence,   120 : 868 = 100 : x,

or,   120 x = 86,800;

whence, $x = 723\frac{1}{3}$ gal.   Ans.

(160) Taking the dimensions as the causes,

$$\begin{array}{c|cc} 14 & 2 & \\ 28 & 20 & \\ 2 & & \\ 12 & 17 & 57 \\ 10 & 6 & \end{array} = 798 \;\middle|\; x, \text{ whence, } 2\,x = 17 \times 57 = 969,$$
or, $x = 484\frac{1}{2}$ bbl.   Ans.

(161) 8 hr. 40 min. = 520 min.   Hence,

$$444 : 1,060 = 520 : x,$$

$$\text{or, } x = \frac{1,060 \times \overset{130}{\cancel{520}}}{\underset{111}{\cancel{444}}} = \frac{137,800}{111} = 1,241.44 + \text{min.} = 20 \text{ hr.}$$
$$41.44 + \text{min. Ans.}$$

(162) 1 min. = 60 sec.   Hence,

$$5\tfrac{1}{2} : 60 = 6,160 : x,$$

$$\text{or, } x = \frac{60 \times 6,160}{5.5} = 67,200 \text{ ft. } \text{Ans.}$$

(163) Writing the statement as a direct proportion, 8 : 10 = 5 : x, it is easy to see that x will be greater than 5; but, it should be smaller, since by working longer hours, fewer men will be required to do the same work. Hence, the proportion is inverse. Inverting the second couplet,

$$8 : 10 = x : 5,$$

$$\text{or, } x = \frac{8 \times \overset{4}{\cancel{5}}}{\underset{2}{\cancel{10}}} = 4 \text{ men. } \textbf{Ans.}$$

(164) Taking the times as the causes,

$$20 \ \Big|\ 25 \ \Big|\ 14$$
$$\Big|\ 5 \ \Big|\ 70$$
$$= 540 \ \Big|\ 630;$$
$$10 \ \Big|\ x \quad 27$$
$$2 \ \Big|\ 3$$

whence, $3x = 2 \times 14 = 28$, or $x = 9\tfrac{1}{3}$ hr.   Ans.

(165) Taking the horsepowers as the effects, we have for the known causes in example 4, Art. **349**, 14', 500, and 48, and for the known effect 112 horsepower. Hence,

$$\begin{array}{c|c}
& 9 \\
14 & 990 \\
198 & 22 \\
14' \ \Big|\ 30'\cdot & 5 \ \Big|\ 110 \\
500 \ \Big|\ 660 = 112 \ \Big|\ x, \text{ or } 500 \ \Big|\ 660 = 112 \ \Big|\ x; \\
48 \ \Big|\ 42 & 6 \ \Big|\ 3 \quad 8 \\
& 48 \ \Big|\ 42
\end{array}$$

whence, $x = 9 \times 22 \times 3 = 594$ horsepower.   Ans.

(166) First find the volume of the cylinder in cubic inches, as in the example, Art. **345**. The volume, multiplied by the weight of one cubic inch (.261 lb.), will evidently be the weight of the cylinder. Hence,

$$10' \ \Big|\ 12' \ \Big|\ = 1{,}570.8 \ \Big|\ x, \text{ or } \begin{array}{c|c} 100 & 144 \\ & 3 \\ 20 & 60 \end{array} = 1.570.8 \ \Big|\ x;$$

whence, $x = \dfrac{144 \times 3 \times 1{,}570.8}{100} = 6{,}785.856$ cu. in. Therefore, weight of cylinder $= 6{,}785.856 \times .261 = 1{,}771.11$ lb.   Ans.

(167) Referring to the example in Art. **348**,

$$\begin{array}{c|c}
& 4 \\
5 & 40 \\
15 \ \Big|\ 40 & 15 \\
& 100 \\
20' \ \Big|\ 18' = 187 \ \Big|\ x, \text{ or } & 400 \ \Big|\ 324 = 187 \ \Big|\ x; \\
10 \ \Big|\ 12 & 10 \ \Big|\ 4 \\
& 12
\end{array}$$

whence, $x = \dfrac{324 \times 4 \times 187}{500} = 484.7$ lb.   Ans.

ALGEBRA.

(QUESTIONS 168–217.)

---

(168) $-\dfrac{c-(a-b)}{c+(a+b)} = \dfrac{(a-b)-c}{c+(a+b)}$. Ans. (Art. **482.**)

(169) (*a*) Factoring each expression (Art. **457**), we have $9x^4 + 12x^2y^2 + 4y^4 = (3x^2 + 2y^2)(3x^2 + 2y^2) = (3x^2 + 2y^2)^2$.

(*b*) $49a^4 - 154a^2b^2 + 121b^4 = (7a^2 - 11b^2)(7a^2 - 11b^2) = (7a^2 - 11b^2)^2$. Ans.

(*c*) $64x^2y^2 + 64xy + 16 = 16(2xy + 1)^2$. Ans.

(170) (*a*) Arrange the dividend according to the decreasing powers of x and divide. Thus,

$$3x - 1) 9x^3 + 3x^2 + x - 1 (3x^2 + 2x + 1 \quad \text{Ans.}$$
$$\underline{9x^3 - 3x^2}$$
$$6x^2 + x \qquad \text{(Art. } \mathbf{444.)}$$
$$\underline{6x^2 - 2x}$$
$$3x - 1$$
$$3x - 1$$

(*b*)
$$a - b) a^3 - 2ab^2 + b^3 (a^2 + ab - b^2 \quad \text{Ans.}$$
$$\underline{a^3 - a^2b}$$
$$a^2b - 2ab^2 \qquad \text{(Art. } \mathbf{444.)}$$
$$\underline{a^2b - ab^2}$$
$$-ab^2 + b^3$$
$$-ab^2 + b^3$$

(*c*) Arranging the terms of the dividend according to the decreasing powers of x, we have

$$7x - 3) 7x^3 - 24x^2 + 58x - 21 (x^2 - 3x + 7 \quad \text{Ans.}$$
$$\underline{7x^3 - 3x^2}$$
$$-21x^2 + 58x$$
$$\underline{-21x^2 + 9x}$$
$$49x - 21$$
$$49x - 21$$

(171) See Arts. **352** and **353.**

(172) (a) In the expression $4x^3y - 12x^2y^2 + 8xy^3$, it is evident that each term contains the common factor $4xy$. Dividing the expression by $4xy$, we obtain $x^2 - 3x^2y + 2y^2$ for a quotient. The two factors, therefore, are $4xy$ and $x^2 - 3x^2y + 2y^2$. Hence, by Art. **452,**

$$4x^3y - 12x^2y^2 + 8xy^3 = 4xy(x^2 - 3x^2y + 2y^2). \quad \text{Ans.}$$

(b) The expression $(x^4 - y^4)$ when factored, equals $(x^2 + y^2)(x^2 - y^2)$. (Art. **463.**) But, according to Art. **463,** $x^2 - y^2$ may be further resolved into the factors $(x + y)(x - y)$.

Hence, $(x^4 - y^4) = (x^2 + y^2)(x + y)(x - y)$. Ans.

(c) $8x^3 - 27y^3$. See Art. **466.** The cube root of the first term is $2x$, and of the second term is $3y$, the sign of the second term being $-$. Hence, the first factor of $8x^3 - 27y^3$ is $2x - 3y$. The second factor we find to be $4x^2 + 6xy + 9y^2$, by division. Hence, the factors are $2x - 3y$ and $4x^2 + 6xy + 9y^2$.

(173). Arranging the terms according to the decreasing powers of m.

$$
\begin{array}{l}
3m^3 + 10m^2n + 10mn^2 + 3n^3 \\
3m^4n - 5m^3n^2 + 5m^2n^3 - mn^4 \\
\hline
9m^7n + 30m^6n^2 + 30m^5n^3 + 9m^4n^4 \\
\quad - 15m^6n^2 - 50m^5n^3 - 50m^4n^4 - 15m^3n^5 \\
\quad\quad + 15m^5n^3 + 50m^4n^4 + 50m^3n^5 + 15m^2n^6 \\
\quad\quad\quad - 3m^4n^4 - 10m^3n^5 - 10m^2n^6 - 3mn^7 \\
\hline
9m^7n + 15m^6n^2 - 5m^5n^3 + 6m^4n^4 + 25m^3n^5 + 5m^2n^6 - 3mn^7
\end{array}
$$

Ans.

(174) $(2a^2bc^3)^4 = 16a^8b^4c^{12}$. Ans.

$(-3a^2b^2c)^5 = -243a^{10}b^{10}c^5$. Ans.

$(-7m^3nx^2y^4)^2 = 49m^6n^2x^4y^8$. Ans.

(175) (a) $4a^2 - b^2$ factored $= (2a + b)(2a - b)$. Ans.

(b) $16x^{10} - 1$ factored $= (4x^5 + 1)(4x^5 - 1)$. (Art. **463.**)

Ans.

(c) $16x^4 - 8x^4y^2 + x^2y^4$, when factored $=$

$(4x^2 - xy^2)(4x^2 - xy^2)$. (Art. **457**, Rule.)

But, $(4x^2 - xy^2) = x(2x + y)(2x - y)$. (Arts. **452** and **463**.)

Hence, $16x^4 - 8x^4y^2 + x^2y^4 = x^2 (2x + y) (2x + y) (2x - y)$ $(2x - y)$. Ans.

(**176**) $\dfrac{4a^4 - 12a^4x + 5a^4x^2 + 6a^2x^3 + a^2x^4(2a^2 - 3a^2x - ax^2}{4a^2}$ Ans.

$$
\begin{array}{l|l}
4a^2 - 3a^2x & -12a^4x + 5a^4x^2 \\
 & -12a^4x + 9a^4x^2 \\ \hline
4a^2 - 6a^2x - ax^2 & -\ \ 4a^4x^2 + 6a^2x^3 + a^2x^4 \\ \hline
 & -\ \ 4a^4x^2 + 6a^2x^3 + a^2x^4
\end{array}
$$

(**177**)   (a) $6a^4b^4 + a^3b^2 - 7a^2b^3 + 2abc + 3$.

(b) $3 + 2abc + a^3b^2 - 7a^2b^3 + 6a^4b^4$.

(c) $1 + ax + a^2 + 2a^3$. Written like this, the a in the second term is understood as having 1 for an exponent; hence, if we represent the first term by a^0, in value it will be equal to 1, since $a^0 = 1$. (Art. **439**.) Therefore, 1 should be written as the first term when arranged according to the increasing powers of a.

(**178**) $\sqrt[4]{16a^{12}b^4c^8} = \pm\, 2a^3bc^2$. Ans. (Art. **521**.)

$\sqrt[5]{-32a^{15}} = -2a^3$. Ans.

$\sqrt[3]{-1,728a^6d^{12}x^3y^6} = -12a^2d^4xy^2$. Ans.

(**179**)   (a) $(a - 2x + 4y) - (3z + 2b - c)$. Ans. (Art. **408**.)

(b) $-3b - 4c + d - (2f - 3e)$ becomes $-[3b + 4c - d + (2f - 3e)]$ when placed in brackets preceded by a minus sign. Ans. (Art. **408**.)

(c) The subtraction of one expression or quantity from another, when none of the terms are alike, can be represented only by combining the subtrahend with the minuend by means of the sign $-$.

In this case, where we are to subtract
$2b - (3c + 2d) - a$ from x, the result will be indicated by
$x - [2b - (3c + 2d) - a.]$ Ans.   (Art. **408**.)

(**180**)   (a) $2x^3 + 2x^2 + 2x - 2$

$\underline{x - 1}$

$2x^4 + 2x^3 + 2x^2 - 2x$

$\underline{- 2x^3 - 2x^2 - 2x + 2}$

$2x^4 - 4x + 2$ Ans.

(b) $x^2 - 4ax + c$

$\underline{2x + a}$

$2x^3 - 8ax^2 + 2cx$

$\underline{ax^2 - 4a^2x + ac}$

$2x^3 - 7ax^2 + 2cx - 4a^2x + ac$ Ans.

(c) $- a^3 + 3a^2b - 2b^3$

$\underline{5a^2 + 9ab}$

$- 5a^5 + 15a^4b - 10a^2b^3$

$\underline{- 9a^4b + 27a^3b^2 - 18ab^4}$

$- 5a^5 + 6a^4b - 10a^2b^3 + 27a^3b^2 - 18ab^4$

Arranging the terms according to the decreasing powers
of a, we have $- 5a^5 + 6a^4b + 27a^3b^2 - 10a^2b^3 - 18ab^4$.   Ans.

(**181**) (a) $4xyz$

$- 3xyz$

$- 5xyz$

$6xyz$

$- 9xyz$

$3xyz$

$\overline{- 4xyz}$ Ans.

The sum of the coefficients
of the positive terms we find to
be $+ 13$, since $(+ 3) + (+ 6) +$
$(+ 4) = (+ 13)$.

When no sign is given before
a quantity the $+$ sign must always be understood.   The sum
of the coefficients of the nega-
tive terms we find to be $- 17$ since $(- 9) + (- 5) + (- 3) =$
$(- 17)$.   Subtracting the *lesser* sum from the *greater*, and
prefixing the sign of the greater sum $(-)$ (Art. **390**, rule
II), we have $(+ 13) + (- 17) = - 4$.   Since the terms are
all alike, we have only to annex the common symbols xyz to
$- 4$, thereby obtaining $- 4xyz$ for the result or sum.

(b) $3a^2 + 2ab + 4b^2$
$\quad\; 5a^2 - 8ab + b^2$
$\quad -a^2 + 5ab - b^2$
$\quad 18a^2 - 20ab - 19b^2$
$\quad 14a^2 - 3ab + 20b^2$
$\quad\overline{39a^2 - 24ab + 5b^2}$ Ans.

When adding polynomials, always place like terms under each other. (Art. **393**.)

The coefficient of a^2 in the result will be 39, since $(+14)+(+18)+(-1)+(+5)+(+3)=39$. When the coefficient of a term is not written, 1 is always understood to be its coefficient. (Art. **359**.) The coefficient of ab will be -24, since $(-3)+(-20)+(+5)+(-8)+(+2)=-24$. The coefficient of b^2 will be $(+20)+(-19)+(-1)+(+1)+(+4)=+5$. Hence, the result or sum is $39a^2 - 24ab + 5b^2$.

(c) $\quad 4mn + 3ab - 4c$
$\quad +2mn - 4ab \quad\quad + 3x + 3m^2 - 4p$
$\quad\overline{6mn - ab - 4c + 3x + 3m^2 - 4p}$ Ans.

(182) The reciprocal of 3.1416 is $\dfrac{1}{3.1416} = .3183+.$ Ans.

Reciprocal of .7854 $= \dfrac{1}{.7854} = 1.273+.$ Ans.

Of $\dfrac{1}{64.32} = \dfrac{1}{\dfrac{1}{64.32}} = 1 \times \dfrac{64.32}{1} = 64.32.$ Ans. (See Art. **481**.)

(183) (a) $\dfrac{x}{x-y} + \dfrac{x-y}{y-x}.$ If the denominator of the second fraction were written $x-y$, instead of $y-x$, then $x-y$ would be the common denominator.

By Art. **482**, the signs of the denominator and the sign before the fraction $\dfrac{x-y}{y-x}$ may be changed, giving $-\dfrac{x-y}{x-y}.$ We now have

$$\dfrac{x}{x-y} - \dfrac{x-y}{x-y} = \dfrac{x-x+y}{x-y} = \dfrac{y}{x-y}.$$ Ans.

(b) $\dfrac{x^2}{x^2-1} + \dfrac{x}{x+1} - \dfrac{x}{1-x}.$ If we write the denominator of the third fraction $x-1$ instead of $1-x$, x^2-1 will then be the common denominator.

By Art. **482,** the signs of the denominator and the sign before the fraction may be changed, thereby giving $\dfrac{x}{x-1}$. We now have

$$\frac{x^2}{x^2-1}+\frac{x}{x+1}+\frac{x}{x-1}=\frac{x^2+x(x-1)+x(x+1)}{x^2-1}-$$

$$\frac{x^2+x^2-x+x^2+x}{x^2-1}=\frac{3x^2}{x^2-1}. \quad \text{Ans.}$$

(c) $\dfrac{3a-4b}{7}-\dfrac{2a-b+c}{3}+\dfrac{13a-4c}{12}$, when reduced to a common denominator

$$=\frac{12(3a-4b)-28(2a-b+c)+7(13a-4c)}{84}.$$

Expanding the terms and removing the parentheses, we have

$$\frac{36a-48b-56a+28b-28c+91a-28c}{84}.$$

Combining like terms in the numerator, we have as the result,

$$\frac{71a-20b-56c}{84}. \quad \text{Ans.}$$

(184)   (a) $45x^7y^{10}-90x^5y^7-360x^4y^8=$
$45x^4y^7(x^3y^3-2x-8y)$.   Ans.   (Art. **452.**)

(b) $a^2b^2+2abcd+c^2d^2=(ab+cd)^2$.   Ans.   (Art. **457.**)

(c) $(a+b)^2-(c-d)^2=(a+b+c-d)(a+b-c+d)$.
Ans.   (Art. **463.**)

(185)   (a) If a man builds 20 rods of stone wall, and we consider this work as positive, or $+$, the work which he does in tearing it down may be considered as negative, or $-$. If he tore down 10 rods, we could say that he *built* -10 rods.

(b)   See Arts. **388** and **398.**

(186)   (a) $\dfrac{2ax+x^2}{a^2-x^2}\div\dfrac{x}{a-x}=\dfrac{x(2a+x)}{a^2-x^2}\times\dfrac{a-x}{x}.$
(Art. **502.**)

Canceling common factors, the result equals $\dfrac{2a + x}{a^2 + ax + x^2}$. Ans.

$$a - x \,)\, a^3 - x^3 \,(\, a^2 + ax + x^2$$
$$\underline{a^3 - a^2x}$$
$$a^2x - x^3$$
$$\underline{a^2x - ax^2}$$
$$ax^2 - x^3$$
$$\underline{ax^2 - x^3}$$

(b) Inverting the divisor and factoring, we have

$$\frac{3n(2m^2n - 1)}{(2m^2n - 1)\,(2m^2n - 1)} \times \frac{(2m^2n + 1)\,(2m^2n - 1)}{3n}.$$

Canceling common factors, we have $2m^2n + 1$. Ans.

(c) $9 + \dfrac{5y^2}{x^2 - y^2} \div \left(3 + \dfrac{5y}{x - y}\right)$ simplified $= \dfrac{9x^2 - 4y^2}{x^2 - y^2} \div \dfrac{3x + 2y}{x - y}$.

Inverting the divisor, we have $\dfrac{9x^2 - 4y^2}{x^2 - y^2} \times \dfrac{x - y}{3x + 2y}$.

Canceling common factors, the result equals $\dfrac{3x - 2y}{x + y}$. Ans.

(187) According to Art. 456, the trinomials $1 - 2x^2$ $+ x^4$ and $4x^2 + 4x + 1$ are perfect squares. (See Art. 458.) The remaining trinomials are not perfect squares, since they do not comply with the foregoing principles.

(188) (a) By Art. 481, the reciprocal of $\dfrac{24}{49} = 1 \div \dfrac{24}{49}$

$= 1 \times \dfrac{49}{24} = \dfrac{49}{24}$. Ans.

(b) Since, by Art. 481, a number may be found from its reciprocal by dividing 1 by the reciprocal, the number $= 1 \div 700 = .0014\frac{2}{7}$. Ans.

(**189**)   Applying the method of Art. **474**,

$x+y$	$12xy\,(x^2-y)$,	$2x^2(x^2+2xy+y^2)$,	$3y^2(x-y)^2$,	$6(x^2+xy)$
$x-y$	$12xy\,(x-y)$,	$2x^2(x+y)$,	$3y^2(x-y)^2$,	$6x$
$3xy$	$12xy$	$2x^2(x+y)$,	$3y^2(x-y)$,	$6x$
2	4,	$2x\,(x+y)$,	$y(x-y)$,	2
	2,	$x\,(x+y)$,	$y(x-y)$,	1

Whence,   L. C. M. $= (x+y)\ (x-y)\ 3\,x\,y \times 2 \times 2 \times x\,(x+y) \times y\,(x-y) = 12x^2y^2\,(x+y)^2\,(x-y)^2$.

(**190**) (a) $2 + 4a - 5a^2 - 6a^3$

$\qquad\quad 7a^2$

$\qquad\overline{14a^2 + 28a^4 - 35a^5 - 42a^6}$ Ans.   (Art. **423**.)

(b) $4x^2 - 4y^2 + 6z^2$

$\qquad 3x^2y$

$\qquad\overline{12x^4y - 12x^2y^3 + 18x^2yz^2}$ Ans.

(c) $3b + 5c - 2d$

$\qquad 6a$

$\qquad\overline{18ab + 30ac - 12ad}$ Ans.

(**191**) (a) See Arts. **359** and **361**.

(b) See Arts. **419** and **440**.

(c) See Art. **416**.

(**192**) (a)   On removing the vinculum, we have
$2a - [3b + \{4c - 4a - (2a + 2b)\} + \{3a - b - c\}\,]$.

$\qquad\qquad\qquad\qquad\qquad\qquad\qquad$ (Art. **405**.)

Removing the parenthesis,
$\quad 2a - [3b + \{4c - 4a - 2a - 2b\} + \{3a - b - c\}\,]$.

Removing the braces,
$\quad\ \ 2a - [3b + 4c - 4a - 2a - 2b + 3a - b - c]$.

$\qquad\qquad\qquad\qquad\qquad\qquad\qquad$ (Art. **406**.)

Removing the brackets,
$\qquad 2a - 3b - 4c + 4a + 2a + 2b - 3a + b + c$.

Combining like terms, the result is $5a - 3c$.   Ans.

(*b*) Removing the parenthesis, we have

$$7a - [3a - \{2a - 5a + 4a\}].$$

Removing the brace,

$$7a - [3a - 2a + 5a - 4a].$$

Removing the brackets,

$$7a - 3a + 2a - 5a + 4a.$$

Combining terms, the result is $5a$. Ans.

(*c*) Removing the parentheses, we have

$$a - [2b + \{3c - 3a - a - b\} + \{2a - b - c\}].$$

Removing the braces,

$$a - [2b + 3c - 3a - a - b + 2a - b - c].$$

Removing the brackets,

$$a - 2b - 3c + 3a + a + b - 2a + b + c.$$

Combining like terms, the result is $3a - 2c$. Ans.

(193) (*a*) $(x^3 + 8) = (x + 2)(x^2 - 2x + 4)$. Ans.

(*b*) $x^3 - 27y^3 = (x - 3y)(x^2 + 3xy + 9y^2)$. Ans.

(*c*) $xm - nm + xy - ny = m(x - n) + y(x - n)$,

or $(x - n)(m + y)$. Ans.

(Arts. **466** and **468**.)

(194) Arrange the terms according to the decreasing powers of x. (Art. **523**.)

$4x^4 + 8ax^3 + 4a^2x^2 + 16b^2x^2 + 16ab^2x + 16b^4 (2x^2 + 2ax + 4b^2$. Ans.
$(2x^2)^2 = 4x^4$.

$4x^2 + 2ax$	$8ax^3 + 4a^2x^2$
	$8ax^3 + 4a^2x^2$
$4x^2 + 4ax + 4b^2$	$16b^2x^2 + 16ab^2x + 16b^4$
	$16b^2x^2 + 16ab^2x + 16b^4$

(195) $\dfrac{c(a + b) + cd}{(a + b)c} = \dfrac{ac + bc + cd}{ac + bc}$. Canceling c, which is common to each term, we have $\dfrac{a + b + d}{a + b} = 1 + \dfrac{d}{a + b}$. Ans.

(196) (*a*) $x + y + z - (x - y) - (y + z) - (-y)$ becomes $x + y + z - x + y - y - z + y$ on the removal of the
parentheses. (Art. **405**.) Combining like terms,

$$x - x + y + y - y + y + z - z = 2y. \quad \text{Ans.}$$

(*b*) $(2x - y + 4z) + (-x - y - 4z) - (3x - 2y - z)$ becomes $2x - y + 4z - x - y - 4z - 3x + 2y + z$, on the
removal of the parentheses. (Arts. **405** and **406**.) Combining like terms,

$$2x - x - 3x - y - y + 2y + 4z - 4z + z = z - 2x. \quad \text{Ans.}$$

(*c*) $a - [2a + (3a - 4a)] - 5a - \{6a - [(7a + 8a) - 9a]\}$.

In this expression we find aggregation marks of different
shapes, thus, [, (, and {. In such cases look for the corresponding part (whatever may intervene), and all that is included between the two parts of each aggregation mark
must be treated as directed by the sign before it (Arts. **405**
and **406**), no attention being given to any of the other aggregation marks. It is always best to begin with the *inner-
most pair*, and remove each pair of aggregation marks in
order. First removing the parentheses, we have

$$a - [2a + 3a - 4a] - 5a - \{6a - [7a + 8a - 9a]\}.$$

Removing the brackets, we have

$$a - 2a - 3a + 4a - 5a - \{6a - 7a - 8a + 9a\}.$$

Removing the brace, we have

$$a - 2a - 3a + 4a - 5a - 6a + 7a + 8a - 9a.$$

Combining like terms, the result is $-5a$. Ans.

(197) (*a*) A square x square, plus $2a$ cube b fifth,
minus the quantity a plus b.

(*b*) The cube root of x, plus y into the quantity a minus
n square to the $\frac{2}{3}$ power.

(*c*) The quantity m plus n, into the quantity m minus
n squared into the quantity m minus the quotient of n
divided by 2.

(198)

$$a^2 - a^2 - 2a - 1)2a^4 - 4a^5 - 5a^4 + 3a^3 + 10a^2 + 7a + 2(2a^3 - 2a^2 - 3a - 2$$
$$2a^4 - 2a^3 - 4a^4 - 2a^3 \qquad\qquad \text{Ans.}$$

$$\overline{ -2a^5 - a^4 + 5a^3 + 10a^2}$$
$$-2a^5 + 2a^4 + 4a^3 + 2a^2$$

$$\overline{ -3a^4 + a^3 + 8a^2 + 7a}$$
$$-3a^4 + 3a^3 + 6a^2 + 3a$$

$$\overline{ -2a^3 + 2a^2 + 4a + 2}$$
$$-2a^3 + 2a^2 + 4a + 2$$

(199) (*a*) Factoring according to Art. **452**, we have $x^2y^2 (x^6 - 64)$. Factoring ($x^6 - 64$), according to Art. **463**, we have

$$(x^3 + 8) (x^3 - 8).$$

Art. **466**, rule. $x^3 + 8 = (x + 2) (x^2 - 2x + 4)$.

Art. **466**, rule. $x^3 - 8 = (x - 2) (x^2 + 2x + 4)$.

Therefore, $x^6y^2 - 64x^2y^2 = x^2y^2 (x + 2) (x^2 - 2x + 4) (x - 2)$ $(x^2 + 2x + 4)$, or $x^2y^2 (x + 2) (x - 2) (x^2 + 2x + 4)$ $(x^2 - 2x + 4)$. Ans.

(*b*) $a^2 - b^2 - c^2 + 1 - 2a + 2bc$. Arrange as follows (Art. **408**):

$$(a^2 - 2a + 1) - (b^2 - 2bc + c^2) = (a - 1)^2 - (b - c)^2.$$
(Art. **455**.)

By Art. **463**, we have

$$(a - 1 + b - c) (a - 1 - [b - c]),$$
$$\text{or} \quad (a - 1 + b - c) (a - 1 - b + c). \quad \text{Ans.}$$

(*c*) $1 - 16a^2 + 8ac - c^2$. Placing the last three terms in parentheses (Art. **408**), $1 - (16a^2 - 8ac + c^2)$.

$16a^2 - 8ac + c^2 = (4a - c)^2$. (Art. **455**.)

$1 - (16a^2 - 8ac + c^2) = 1 - (4a - c)^2$.

$1 - (4a - c)^2 = [1 + (4a - c)] [1 - (4a - c)]$. (Art. **463**.)

Removing parentheses, and writing parentheses in place of the brackets,

$$1 - (4a - c)^2 = (1 + 4a - c) (1 - 4a + c). \quad \text{Ans.}$$

(200) See Art. **482**.

(201) The subtraction of one expression from another, if none of the terms are similar, may be represented only by connecting the subtrahend with the minuend by means of the sign —. Thus, it is required to subtract $5a^3b - 7a^2b^2 + 5ab^3$ from $a^4 - b^4$, the result will be represented by $a^4 - b^4 - (5a^3b - 7a^2b^2 + 5ab^3)$, which, on removing the parentheses (Art. **405**), becomes $a^4 - b^4 - 5a^3b + 7a^2b^2 - 5ab^3$. From this result, subtract $3a^4 - 4a^3b + 6a^2b^2 + 5ab^3 - 3b^4$.

$\quad\quad a^4 - \quad b^4 - 5a^3b + 7a^2b^2 - 5ab^3$ *minuend*.

$\underline{\;-3a^4 + 3b^4 + 4a^3b - 6a^2b^2 - 5ab^3}$ *subtrahend*, with signs

$\quad -2a^4 + 2b^4 - \quad a^3b + a^2b^2 - 10ab^3$ changed. (Art. **401**.)

Or, $-2a^4 - a^3b + a^2b^2 - 10ab^3 + 2b^4$. Answer arranged according to the decreasing powers of a.

(202)   (a) $3a - 2b + 3c$ $3a - 2b + 3c$

$\quad\quad\quad\quad\quad \underline{2a - 8b - \;c}$ becomes $\underline{-2a + 8b + \;c}$

$\quad\quad\quad\quad\quad\quad\quad\quad\quad\quad\quad\quad\quad\quad\quad\quad\quad a + 6b + 4c$

when the signs of the subtrahend are changed. Now, adding each term (with its sign changed) in the subtrahend to its corresponding term in the minuend, we have $(-2a) + (3a) = a$; $(+8b) + (-2b) = +6b$; $(+c) + (3c) = +4c$. Hence, $a + 6b + 4c$ equals the difference. Ans.

(b) $2x^3 - 3x^2y + 2xy^2$

$\quad\quad x^3 \quad\quad\quad\quad\quad\quad\quad + y^3 - xy^2$ becomes

$\quad\quad\quad\quad\quad\quad \overline{2x^3 - 3x^2y + 2xy^2}$

$\quad\quad\quad\quad\quad\quad -\;x^3 \quad\quad\quad\quad\quad -y^3 + xy^2$

$\quad\quad\quad\quad\quad\quad \overline{x^3 - 3x^2y + 2xy^2 - y^3 + xy^2}$

when the signs of the subtrahend are changed. Adding each term in the subtrahend (with its sign changed) to its corresponding term in the minuend, we have $x^3 - 3x^2y + 2xy^2 - y^3 + xy^2$, which, arranged according to the decreasing powers of x, equals $x^3 - 3x^2y + xy^2 + 2xy^2 - y^3$. Ans.

(c) $14a + 4b - 6c - 3d$

$\quad\quad\quad\quad\quad\quad\quad\quad 11a - 2b + 4c - 4d$

On changing the sign of each term in the subtrahend, the problem becomes

$$14a + 4b - 6c - 3d$$
$$- 11a + 2b - 4c + 4d$$
$$\overline{\quad 3a + 6b - 10c + d \quad}$$

Adding each term of the subtrahend (with the sign changed) to its corresponding term in the minuend, the difference, or result, is $3a + 6b - 10c + d$. Ans.

(**203**) The numerical values of the following, when $a = 16$, $b = 10$, and $x = 5$, are:

(*a*) $(ab^2x + 2abx)\, 4a = (16 \times 10^2 \times 5 + 2 \times 16 \times 10 \times 5) \times 4 \times 16$. It must be remembered that when no sign is expressed between symbols or quantities, the sign of multiplication is understood.

$(16 \times 100 \times 5 + 2 \times 16 \times 10 \times 5) \times 64 = (8,000 + 1,600) \times 64 = 9,600 \times 64 = 614,400$. Ans.

(*b*) $2\sqrt{4a} - \dfrac{2bx}{a-b} + \dfrac{b-x}{x} = 2\sqrt{64} - \dfrac{2 \times 10 \times 5}{16-10} +$

$\dfrac{10-5}{5} = 16 - \dfrac{100}{6} + 1 = \dfrac{96 - 100 + 6}{6} = \dfrac{2}{6} = \dfrac{1}{3}$. Ans.

(*c*) $(b - \sqrt{a})\,(x^3 - b^2)\,(a^2 - b^2) = (10 - \sqrt{16})\,(5^3 - 10^2) (16^2 - 10^2) = (10 - 4)\,(125 - 100)\,(256 - 100) = 6 \times 25 \times 156 = 23,400$. Ans.

(**204**) (*a*) Dividing both numerator and denominator by $15\,m\,x\,y^2$, $\dfrac{15\,m\,x\,y^2}{75\,m\,x^2y^3} = \dfrac{1}{5\,x\,y}$. Ans.

(*b*) $\dfrac{x^2 - 1}{4\,x\,(x+1)} = \dfrac{(x+1)\,(x-1)}{4\,x\,(x+1)}$ when the numerator is factored.

Canceling $(x + 1)$ from both the numerator and denominator (Art. **484**), the result is $\dfrac{x-1}{4\,x}$. Ans.

(*c*) $\dfrac{(a^3 + b^3)\,(a^3 + a\,b + b^3)}{(a^3 - b^3)\,(a^3 - a\,b + b^3)}$ when factored becomes

$\dfrac{(a+b)\,(a^2 - a\,b + b^2)\,(a^2 + a\,b + b^2)}{(a-b)\,(a^2 + a\,b + b^2)\,(a^2 - a\,b + b^2)}$. (Art. **466**.)

Canceling the factors common to both the numerator and denominator, we have

$$\frac{(a+b)(a^2-ab+b^2)(a^2+ab+b^2)}{(a-b)(a^2+ab+b^2)(a^2-ab+b^2)} = \frac{a+b}{a-b}. \quad \text{Ans.}$$

$$a+b\ )\ a^3+b^3\ (\ a^2-ab+b^2 \qquad a-b\ )\ a^3-b^3\ (\ a^2+ab+b^2$$

$$\underline{a^3+a^2b} \qquad\qquad\qquad \underline{a^3-a^2b}$$

$$-a^2b+b^3 \qquad\qquad\qquad a^2b-b^3$$
$$\underline{-a^2b-ab^2} \qquad\qquad\qquad \underline{a^2b-ab^2}$$

$$\underline{\quad ab^2+b^3\quad} \qquad\qquad\qquad \underline{\quad ab^2-b^3\quad}$$
$$ab^2+b^3 \qquad\qquad\qquad\qquad ab^2-b^3$$

(205) (a)

$$\frac{\dfrac{1}{1-x}-\dfrac{1}{1+x}}{\dfrac{1}{1-x}+\dfrac{1}{1+x}} = \frac{\dfrac{1+x-1+x}{1-x^2}}{\dfrac{1+x+1-x}{1-x^2}} = \frac{2x}{1-x^2} \div$$

$$\frac{2}{1-x^2} = \frac{2x}{1-x^2} \times \frac{1-x^2}{2} = x. \quad \text{Ans. (See Art. \textbf{509}.)}$$

(b)

$$\frac{\dfrac{a^2}{b^3}+\dfrac{1}{a}}{\dfrac{a}{b^3}-\dfrac{a-b}{ab}} = \frac{\dfrac{a^3+b^3}{ab^3}}{\dfrac{a^2b-b^2(a-b)}{ab^3}} = \frac{\dfrac{a^3+b^3}{ab^3}}{\dfrac{a^2b-ab^2+b^3}{ab^3}} =$$

$$\frac{a^3+b^3}{ab^3} \div \frac{a^2b-ab^2+b^3}{ab^3} = \frac{a^3+b^3}{ab^3} \times \frac{ab^3}{b(a^2-ab+b^2)} = \frac{a+b}{b}.$$
$$\text{Ans.}$$

(c)

$$\cfrac{1}{x+\cfrac{1}{1+\dfrac{x+1}{3-x}}} = \cfrac{1}{x+\dfrac{3-x}{4}} = \frac{4}{3x+3}. \quad \text{Ans.}$$
$$\text{(Art. \textbf{509}.)}$$

(206) $\dfrac{3+2x}{2-x} - \dfrac{2-3x}{2+x} + \dfrac{16x-x^2}{x^2-4}$. If the denominator of the third fraction were written $4-x^2$, instead of x^2-4, the common denominator would then be $4-x^2$.

By Art. **482**, $\dfrac{16x-x^2}{-4}$ becomes $-\dfrac{16x-x^2}{-x^2+4} = -\dfrac{16x-x^2}{4-x^2}$.

Hence, $\dfrac{3+2x}{2-x} - \dfrac{2-3x}{2+x} - \dfrac{16x-x^2}{4-x^2}$, when reduced to a common denominator, becomes

$$\frac{(3 + 2x)(2 + x) - (2 - 3x)(2 - x) - (16x - x^2)}{4 - x^2} =$$

$$\frac{(6 + 7x + 2x^2) - (4 - 8x + 3x^2) - (16x - x^2)}{4 - x^2}.$$

Removing the parentheses (Art. **405**), we have

$$\frac{6 + 7x + 2x^2 - 4 + 8x - 3x^2 - 16x + x^2}{4 - x^2}.$$

Combining like terms in the numerator, we have

$$\frac{2 - x}{4 - x^2}.$$

Factoring the denominator by Art. **463,** we have

$$\frac{2 - x}{(2 + x)(2 - x)}.$$

Canceling the common factor $(2 - x)$, the result equals

$$\frac{1}{2 + x}, \text{ or } \frac{1}{x + 2}. \quad \text{Ans.} \qquad (\text{Art. } \mathbf{373.})$$

(207) (a)

$$1 + 2x - \frac{4x - 4}{5x} = \frac{5x + 10x^2 - 4x + 4}{5x} = \frac{10x^2 + x + 4}{5x}. \quad \text{Ans.}$$

$$\qquad\qquad\qquad\qquad\qquad\qquad\qquad\qquad (\text{Art. } \mathbf{504.})$$

(b) $\dfrac{3x^2 + 2x + 1}{x + 4} = 3x - 10 + \dfrac{41}{x + 4}$. Ans. (Art. **505.**)

$$x + 4\ )\ 3x^2 + 2x + 1\ (\ 3x - 10 + \frac{41}{x + 4}$$
$$\underline{3x^2 + 12x}$$
$$-10x + 1$$
$$\underline{-10x - 40}$$
$$41$$

(c) Reducing, the problem becomes

$$\frac{x^2 + 4x - 5}{x^2} \times \frac{x - 7}{x^2 - 8x + 7}.$$

Factoring, we have

$$\frac{(x + 5)\ (x - 1)}{x^2} \times \frac{x - 7}{(x - 1)\ (x - 7)}.$$

Canceling common factors, the result equals $\dfrac{x + 5}{x^2}$. Ans.

(**208**) (*a*) Writing the work as follows, and canceling common factors in both numerator and denominator (Arts. **496** and **497**), we have

$$\frac{9m^2n^2}{8p^2q^2} \times \frac{5p^2q}{2xy} \times \frac{24x^2y^2}{90mn} =$$

$$\frac{9 \times 5 \times 24 \times m^2n^2p^2qx^2y^2}{8 \times 2 \times 90 \times m \times n \times p^2 \times q^2 \times x \times y} = \frac{3mnxy}{4pq^2}. \quad \text{Ans.}$$

(*b*) Factoring the numerators and denominators of the fraction (Art. **498**), and writing the factors of the numerators over the factors of the denominators, we have

$$\frac{(a-x)(a^2+ax+x^2)(a+x)(a+x)}{(a+x)(a^2-ax+x^2)(a-x)(a-x)} =$$

$$\frac{(a+x)(a^2+ax+x^2)}{(a-x)(a^2-ax+x^2)}. \quad \text{Ans.}$$

(*c*) This problem may be written as follows, according to Art. **480**:

$$\frac{3ax+4}{1} \times \frac{a^2}{a(3ax+4)(3ax+4)}.$$

Canceling a and $(3ax+4)$, we have $\dfrac{a}{3ax+4}$. Ans.

(**209**) (*a*) $-7my \overline{)\,35m^3y + 28m^2y^2 - 14my^3}$
$$\qquad\qquad -5m^2 - 4my + 2y^2$$
Ans. (Art. **442**.)

(*b*) $a^4 \overline{)\,4a^4 - 3a^2b - a^6b^2}$
$$\qquad 4 - 3ab - a^2b^2 \quad \text{Ans.}$$

(*c*) $4x^2 \overline{)\,4x^3 - 8x^5 + 12x^7 - 16x^9}$
$$\qquad x - 2x^3 + 3x^5 - 4x^7 \quad \text{Ans.}$$

(**210**) (*a*) $16a^2b^2$; a^2+4ab; $4a^2 - 16a^3b + 5a^4 + 7ax$.

(*b*) Since the terms are not alike, we can only indicate the sum, connecting the terms by their proper signs. (Art. **389**.)

(*c*) Multiplication: $4ac^2d$ means $4 \times a \times c^2 \times d$. (Art. **358**.)

(**211**) $\dfrac{a^2 + c^2 + ac}{a^2 + b^2 - c^2 - 2ab} \times \dfrac{a^2 + c^2 - b^2 - 2ac}{a^4c - ac^4}.$

Arranging the terms, we have

$$\frac{a^2 + ac + c^2}{a^2 - 2ab + b^2 - c^2} \times \frac{a^2 - 2ac + c^2 - b^2}{a^4c - ac^4},$$

which, being placed in parentheses, become

$$\frac{a^2 + ac + c^2}{(a^2 - 2ab + b^2) - c^2} \times \frac{(a^2 - 2ac + c^2) - b^2}{a^4c - ac^4}.$$

By Art. **456,** we know that $a^2 - 2ab + b^2$, also $a^2 - 2ac + c^2$, are perfect squares, and may be written $(a - b)^2$ and $(a - c)^2$.

Factoring $a^4c - ac^4$ by Case I, Art. **452,** we have

$$\frac{a^2 + ac + c^2}{(a - b)^2 - c^2} \times \frac{(a - c)^2 - b^2}{ac(a^3 - c^3)} =$$

$$\frac{a^2 + ac + c^2}{(a - b - c)(a - b + c)} \times \frac{(a - c - b)(a - c + b)}{ac(a - c)(a^2 + ac + c^2)}.$$

(Arts. **463** and **466.**)

Canceling common factors and multiplying, we have

$$\frac{a - c + b}{(a - b + c)\,ac\,(a - c)}, \text{ or } \frac{a + b - c}{ac\,(a - b + c)\,(a - c)}. \quad \text{Ans.}$$

(212) The square root of the fraction a plus b plus c divided by n, plus the square root of a, plus the fraction b plus c divided by n, plus the square root of a plus b, plus the fraction c divided by n, plus the quantity a plus b, into c, plus a plus bc.

(213) (a) $\dfrac{4x + 5}{3} - \dfrac{3x - 7}{5x} + \dfrac{9}{12x^2}.$

We will first reduce the fractions to a common denominator. The L. C. M. of the denominator is $60x^2$, since this is the smallest quantity that each denominator will divide without a remainder. Dividing $60x^2$ by 3, the first denominator, the quotient is $20x^2$; dividing $60x^2$ by $5x$, the second denominator, the quotient is $12x$; dividing $60x^2$ by $12x^2$, the third denominator, the quotient is 5. Multiplying the corresponding numerators by these respective quotients, we obtain $20x^2(4x + 5)$ for the first new numerator; $12x(3x - 7)$ for the second new numerator, and $5 \times 9 = 45$ for the third new numerator. Placing these new numerators over the common denominator and expanding the terms, we have

$$\frac{20x^2(4x+5)-12x(3x-7)+45}{60x^2}=\frac{80x^3+100x^2-36x^2+84x+45}{60x^2}.$$

Collecting like terms, the result is

$$\frac{80x^3+64x^2+84x+45}{60x^2}. \quad \text{Ans.}$$

(*b*)　In $\dfrac{1}{2a(a+x)}+\dfrac{1}{2a(a-x)}$, the L. C. M. of the denominators is $2a(a^2-x^2)$, since this is the smallest quantity that each denominator will divide without a remainder. Dividing $2a(a^2-x^2)$ by $2a(a+x)$, the first denominator, we will have $a-x$; dividing $2a(a^2-x^2)$ by $2a(a-x)$, the second denominator, we have $a+x$. Multiplying the corresponding numerators by these respective .quotients, we have $(a-x)$ for the first new numerator, and $(a+x)$ for the second new numerator. Arranging the work as follows:

$$1\times(a-x)=a-x=\text{1st numerator.}$$
$$1\times(a+x)=a+x=\text{2d numerator.}$$
$$\text{or } 2a \qquad = \text{the sum of the numerators.}$$

Placing the $2a$ over the common denominator $2a(a^2-x^2)$, we find the value of the fraction to be

$$\frac{2a}{2a(a^2-x^2)}=\frac{1}{a^2-x^2}. \quad \text{Ans.}$$

(*c*)　$\dfrac{x}{y}+\dfrac{y}{x+y}+\dfrac{x^2}{x^2+xy}=\dfrac{x}{y}+\dfrac{y}{x+y}+\dfrac{x}{x+y}.$

The common denominator $=y(x+y)$. Reducing the fractions to a common denominator, we have

$$\frac{x(x+y)+y^2+xy}{y(x+y)}=\frac{x^2+2xy+y^2}{y(x+y)}=\frac{x+y}{y}. \quad \text{Ans.}$$

(214)　(*a*)　Apply the method of Art. **474:**

$6ax$	$18ax^3,\ 72ay^3,\ 12xy$
$2y$	$3x,\quad 12y^2,\quad 2y$
3	$3x,\quad 6y,\quad 1$
	$x,\quad 2y,\quad 1$

Whence, $6ax \times 2y \times 3 \times x \times 2y = 72ax^2y^2.$　Ans.

(b) $2(1+x) | 4(1+x), 4(1-x), 2(1-x^2)$

$2(1-x) | 2, \quad 2(1-x), \quad 1-x$

$1, \quad\quad 1, \quad\quad 1$

Hence, L. C. M. $= 2(1+x) \times 2(1-x) = 4(1-x^2)$. Ans.

(c) $a-b | (a-b)(b-c), (b-c)(c-a), (c-a)(a-b)$

$b-c | \quad (b-c), (b-c)(c-a), (c-a)$

$c-a | \quad\quad 1, \quad\quad c-a, \quad c-a$

$1, \quad\quad 1, \quad\quad 1$

Hence, L. C. M. $= (a-b)(b-c)(c-a)$. Ans.

(215) $3x^6 - 3 + a - ax^6 = (3-a)x^6 - 3 + a = (3-a)(x^6-1)$. Regarding $x^6 - 1$ as $(x^3)^2 - 1$, we have, by Art. **462**, $x^6 - 1 = (x^3)^2 - 1 = (x^3-1)(x^3+1)$. $x^3 - 1 = (x-1)(x^2+x+1)$; $x^3+1 = (x+1)(x^2-x+1)$. Art. **466.**

Hence, the factors are

$(x^2+x+1)(x^2-x+1)(x+1)(x-1)(3-a)$. Ans.

(216) Arranging the terms according to the decreasing powers of x, and extracting the square root, we have

$x^4 + x^3y + 4\frac{1}{4}x^2y^2 + 2xy^3 + 4y^4 (x^2 + \frac{1}{2}xy + 2y^2$. Ans.

x^4

$2x^2 + \frac{1}{2}xy | x^3y + 4\frac{1}{4}x^2y^2$

$| x^3y + \frac{1}{4}x^2y^2$

$2x^2 + xy + 2y^2 | 4x^2y^2 + 2xy^3 + 4y^4$

$| 4x^2y^2 + 2xy^3 + 4y^4$

(217) The arithmetic ratio of $x^4 - 1$ to $x + 1$ is $x^4 - 1 - (x+1) = x^4 - x - 2$. Art. **381.**

The geometric ratio of $x^4 - 1$ to $x + 1$ is $\dfrac{x^4-1}{x+1} = x^3 - x^2 + x - 1 = (x^2+1)(x-1)$. Ans.

ALGEBRA.

(218) (*a*) According to Art. **528,** $x^{\frac{2}{3}}$ expressed radically is $\sqrt[3]{x^2}$;

$3x^{\frac{1}{2}}y^{-\frac{2}{3}}$ expressed radically is $3\sqrt[6]{x^3y^{-4}}$;

$3x^{\frac{1}{2}}y^{\frac{2}{3}}z^{\frac{1}{2}} = 3\sqrt[6]{x^3y^4z^3}$, since $z^{\frac{1}{2}} = z^{\frac{3}{6}}$. Ans.

(*b*) $a^{-1}b^{\frac{1}{2}} + \dfrac{c^{-2}}{a+b} + (m-n)^{-1} - \dfrac{a^2b^{-2}c}{c^{-3}} =$

$\dfrac{b^{\frac{1}{2}}}{a} + \dfrac{1}{c^2(a+b)} + \dfrac{1}{m-n} - \dfrac{a^2c^4}{b^2}$. Ans.

(*c*) $\sqrt[7]{x^6} = x^{\frac{6}{7}}$. Ans. $\sqrt[3]{x^{-1}} = x^{-\frac{1}{3}}$. Ans.

$(\sqrt[4]{b^6x^3})^2 = (b^{\frac{3}{2}}x^{\frac{3}{4}})^2 = b^{\frac{12}{4}}x^{\frac{3}{2}}$. Ans.

(219) $3\sqrt{21} = \sqrt{189}$. Ans. (Art. **542.**)

$a^2b\sqrt{b^3c} = \sqrt{a^4b^5c}$. Ans. $2x\sqrt[5]{x} = \sqrt[5]{32x^6}$. Ans.

(220) Let $x =$ the length of the post.

Then, $\dfrac{x}{5} =$ the amount in the earth.

$\dfrac{3x}{7} =$ the amount in the water.

$\dfrac{x}{5} + \dfrac{3x}{7} + 13 = x.$

$7x + 15x + 455 = 35x.$

$-13x = -455.$

$x = 35$ feet. Ans.

(221) $t = \dfrac{W_1s_1t_1 + W_2s_2t_2}{W_1s_1 + W_2s_2}.$

In order to transform this formula so that t_2 may stand alone in the first member, we must first clear of fractions. Clearing of fractions, we have

$$t\, W_1\, s_1 + t\, W_2\, s_2 = W_1\, s_1\, t_1 + W_2\, s_2\, t_2.$$

Transposing, we have

$$- W_2\, s_2\, t_2 = W_1\, s_1\, t_1 - t\, W_1\, s_1 - t\, W_2\, s_2.$$

Factoring (Arts. **452** and **408**), we have

$$- W_2\, s_2\, t_2 = W_1\, s_1\, t_1 - (W_1\, s_1 + W_2\, s_2)\, t;$$

whence, $t_2 = \dfrac{(W_1 s_1 + W_2 s_2)\, t - W_1 s_1 t_1}{W_2 s_2}$.   Ans.

(**222**)   Let x = number of miles he traveled per hour.

Then, $\dfrac{48}{x}$ = time it took him.

$\dfrac{48}{x+4}$ = time it would take him if he traveled 4 miles more per hour.

In the latter case the time would have been 6 hours *less;* whence, the equation

$$\frac{48}{x+4} = \frac{48}{x} - 6.$$

Clearing of fractions,

$$48x = 48x + 192 - 6x^2 - 24x.$$

Combining like terms and transposing,

$$6x^2 + 24x = 192.$$

Dividing by 6, $x^2 + 4x = 32.$

Completing the square, $x^2 + 4x + 4 = 36.$

Extracting square root, $x + 2 = \pm 6;$

whence, $x = -2 + 6 = 4,$ or   the number of miles he traveled per hour.   Ans.

(**223**)   (*a*) $S = \sqrt[3]{\dfrac{CPD^2}{f\left(2+\dfrac{D^2}{d^2}\right)}} = \sqrt[3]{\dfrac{CPD^2}{2f+\dfrac{fD^2}{d^2}}}$

Cubing both members to remove the radical,

$$S^3 = \frac{CPD^2}{2f+\dfrac{fD^2}{d^2}}.$$

Simplifying the result, $S^3 = \dfrac{CPD^2 d^2}{2fd^2 + fD^2}.$

Clearing of fractions,
$$2\,S^2fd^2 + S^2fD^2 = CPD^2d^2.$$
Transposing, $CPD^2d^2 = 2\,S^2fd^2 + S^2fD^2$;

whence, $P = \dfrac{2\,S^2fd^2 + S^2fD^2}{CD^2d^2} = \dfrac{(2d^2 + D^2)\,fS^2}{CD^2d^2}.$ Ans.

(*b*) Substituting the values of the letters in the given formula, we have

$$P = \frac{(2 \times 18^2 + 30^2) \times 864 \times 6^2}{10 \times 30^2 \times 18^2} = \frac{(648 + 900) \times 864 \times 216}{9,000 \times 324} =$$

$$\frac{288,893,952}{2,916,000} = 99.1, \text{ nearly. Ans.}$$

(**224**) (*a*) $3x + 6 - 2x = 7x$. Transposing 6 to the second member, and $7x$ to the first member (Art. **561**),
$$3x - 2x - 7x = -6.$$
Combining like terms, $\qquad -6x = -6;$
whence, $\qquad x = 1.$ Ans.

(*b*) $\qquad 5x - (3x - 7) = 4x - (6x - 35).$
Removing the parentheses (Art. **405**),
$$5x - 3x + 7 = 4x - 6x + 35.$$
Transposing 7 to the second member, and $4x$ and $-6x$ to the first member, $5x - 3x - 4x + 6x = 35 - 7.$ (Art. **561**.)
Combining like terms, $\qquad 4x = 28;$
whence, $\qquad x = 28 \div 4 = 7.$ Ans.

(*c*) $\qquad (x + 5)^2 - (4 - x)^2 = 21x.$
Performing the operations indicated, the equation becomes
$$x^2 + 10x + 25 - 16 + 8x - x^2 = 21x.$$
Transposing, $x^2 - x^2 + 10x + 8x - 21x = 16 - 25.$
Combining like terms, $\qquad -3x = -9.$
Dividing by -3, $\qquad x = 3.$ Ans.

(**225**) (*a*) Simplifying by Art. **538**,
$$\sqrt{27} = \sqrt{9} \times \sqrt{3} = 3\sqrt{3}.$$
$$2\sqrt{48} = 2\sqrt{16} \times \sqrt{3} = 8\sqrt{3}.$$
$$3\sqrt{108} = 3\sqrt{36} \times \sqrt{3} = 18\sqrt{3}.$$
Sum $= 29\sqrt{3}.$ Ans. (Art. **544**.)

(b) $\quad \sqrt[3]{128} = \sqrt[3]{64} \times \sqrt[3]{2} = 4\sqrt[3]{2}.$

$\quad\quad \sqrt[3]{686} = \sqrt[3]{343} \times \sqrt[3]{2} = 7\sqrt[3]{2}.$

$\quad\quad \sqrt[3]{16} = \sqrt[3]{8} \times \sqrt[3]{2} = 2\sqrt[3]{2}.$

Sum $= 13\sqrt[3]{2}.$ Ans. (Art. **544.**)

(c) $\sqrt{\dfrac{3}{8}} = \sqrt{\dfrac{3}{8} \times \dfrac{2}{2}} = \sqrt{\dfrac{6}{16}} = \dfrac{1}{4}\sqrt{6}.$ (Art. **540.**)

$\sqrt{\dfrac{1}{6}} = \sqrt{\dfrac{1}{6} \times \dfrac{6}{6}} = \sqrt{\dfrac{6}{36}} = \dfrac{1}{6}\sqrt{6}.$

$\sqrt{\dfrac{2}{27}} = \sqrt{\dfrac{2}{27} \times \dfrac{3}{3}} = \sqrt{\dfrac{6}{81}} = \dfrac{1}{9}\sqrt{6}.$

Sum $= \left(\dfrac{1}{4} + \dfrac{1}{6} + \dfrac{1}{9}\right)\sqrt{6} = \dfrac{19}{36}\sqrt{6}.$ Ans.

(226) Let $x =$ the capacity.

Then, $x - 42 =$ amount held at first;

$\quad\quad 7(x - 42) = x;$

$\quad\quad 7x - 294 = x;$

$\quad\quad\quad 6x = 294;$

$\quad\quad\quad x = 49$ gallons. Ans.

(227) (a) $\quad 2\sqrt{3x + 4} - x = 4.$

Transposing, Art. **579,** so that the radical stands alone in the first member, $\quad 2\sqrt{3x + 4} = x + 4.$

Squaring both members, since the index of the radical is understood to be 2, $\quad 4(3x + 4) = (x + 4)^2,$

$\quad\quad$ or $\quad 12x + 16 = x^2 + 8x + 16.$

Transposing and uniting terms,

$\quad\quad\quad - x^2 - 8x + 12x = 16 - 16.$

$\quad\quad\quad\quad - x^2 + 4x = 0.$

Dividing by $- x,$ $\quad\quad x - 4 = 0;$

$\quad\quad\quad$ whence, $\quad x = 4.$ Ans.

(b) $\quad\quad \sqrt{3x - 2} = 2(x - 4).$

Squaring, $\quad\quad\quad 3x - 2 = 4(x - 4)^2,$

$\quad\quad$ or $\quad 3x - 2 = 4x^2 - 32x + 64.$

Transposing, $-4x^2 + 32x + 3x = 64 + 2.$

Combining terms, $-4x^2 + 35x = 66.$

Dividing by -4, $\quad x^2 - \dfrac{35x}{4} = -\dfrac{66}{4}.$

Completing the square,

$$x^2 - \frac{35x}{4} + \left(\frac{35}{8}\right)^2 = -\frac{66}{4} + \frac{1,225}{64}.$$

$$x^2 - \frac{35x}{4} + \left(\frac{35}{8}\right)^2 = -\frac{1,056}{64} + \frac{1,225}{64} = \frac{169}{64}.$$

Extracting the square root,

$$x - \frac{35}{8} = \pm\frac{13}{8}.$$

Transposing, $\qquad x = \dfrac{35}{8} \pm \dfrac{13}{8} = 6,$ or $2\dfrac{3}{4}.$ Ans.

(c) $\sqrt{x+16} = 2 + \sqrt{x}$ becomes $x + 16 = 4 + 4\sqrt{x} + x,$ when squared.   Canceling x (Art. **562**), and transposing,

$$-4\sqrt{x} = 4 - 16.$$
$$-4\sqrt{x} = -12.$$
$$\sqrt{x} = 3;$$

whence, $x = 3^2 = 9.$ Ans.

(228) (a) $\qquad \sqrt{3x-5} = \dfrac{\sqrt{7x^2 + 36x}}{x}.$

Clearing of fractions,

$$x\sqrt{3x-5} = \sqrt{7x^2 + 36x}.$$

Removing radicals by squaring,

$$x^2(3x - 5) = 7x^2 + 36x.$$
$$3x^3 - 5x^2 = 7x^2 + 36x.$$

Dividing by x, $3x^2 - 5x = 7x + 36.$

Transposing and uniting,

$$3x^2 - 12x = 36.$$

Dividing by 3, $x^2 - 4x = 12.$

Completing the square,
$$x^2 - 4x + 4 = 16.$$
Extracting the square root,
$$x - 2 = \pm 4;$$
whence, $x = 6$, or -2.   Ans.

(b) $x^2 - (b - a)c = ax - bx + cx.$

Transposing, $x^2 - ax + bx - cx = (b - a)c.$

Factoring, $x^2 - (a - b + c)x = bc - ac.$

Regarding $(a - b + c)$ as the coefficient of x, and completing the square,
$$x^2 - (a - b + c)x + \left(\frac{a - b + c}{2}\right)^2 = bc - ac + \left(\frac{a - b + c}{2}\right)^2.$$
$$x^2 - (a - b + c)x + \left(\frac{a - b + c}{2}\right)^2 =$$
$$\frac{a^2 - 2ab + b^2 - 2ac + 2bc + c^2.}{4}$$
$$x - \frac{a - b + c}{2} = \pm \frac{a - b - c}{2}.$$
$$x = \frac{2a - 2b}{2}, \text{ or } \frac{2c}{2}.$$
$$x = a - b, \text{ or } c.   \text{ Ans.}$$

(c) $(x - 2)(x - 4) - 2(x - 1)(x - 3) = 0$, becomes
$$x^2 - 6x + 8 - 2x^2 + 8x - 6 = 0, \text{ when expanded.}$$
Transposing and uniting terms,
$$-x^2 + 2x = -2.$$
Changing signs, $x^2 - 2x = 2.$

Completing the square, $x^2 - 2x + 1 = 3.$

Extracting the square root, $x - 1 = \pm \sqrt{3};$

whence, $x = 1 \pm \sqrt{3}.$ Ans.

(229) (a) $\sqrt{x - 4ab} = \dfrac{(a + b)(a - b)}{\sqrt{x}}.$

Expanding and clearing of fractions,
$$\sqrt{x^2 - 4abx} = a^2 - b^2.$$
Squaring both members,
$$x^2 - 4abx = a^4 - 2a^2b^2 + b^4.$$

Completing the square,

$$x^2 - 4abx + 4a^2b^2 = a^4 + 2a^2b^2 + b^4.$$
$$x - 2ab = \pm (a^2 + b^2).$$
$$x = (a^2 + 2ab + b^2),$$
$$\text{or} - (a^2 - 2ab + b^2).$$
$$x = (a + b)^2, \text{ or } - (a - b)^2. \text{ Ans.}$$

(b) $\quad -\dfrac{1}{\sqrt{x+1}} + \dfrac{1}{\sqrt{x-1}} = \dfrac{1}{\sqrt{x^2-1}}$ becomes

$$-\sqrt{x-1} + \sqrt{x+1} = 1 \text{ when cleared of fractions.}$$

Squaring,

$$x - 1 - 2\sqrt{x^2-1} + x + 1 = 1.$$
$$-2\sqrt{x^2-1} = 1 - 2x.$$

Squaring again, $\quad 4x^2 - 4 = 1 - 4x + 4x^2.$

Canceling $4x^2$ and transposing,

$$4x = 5.$$
$$x = \frac{5}{4} = 1\frac{1}{4}. \quad \text{Ans.}$$

(230) $\quad\quad 5x - 2y = 51.$ $\quad$ (1)

$$19x - 3y = 180. \quad (2)$$

We will first find the value of x by transposing $- 2y$ to the second member of equation (1), whence $5x = 51 + 2y$, and

$$x = \frac{51 + 2y}{5}. \quad (3)$$

This gives the value of x in terms of y. Substituting the value of x for the x in (2), (Art. **609.**)

$$\frac{19(51 + 2y)}{5} - 3y = 180.$$

Expanding, $\quad\quad\quad \dfrac{969 + 38y}{5} - 3y = 180.$

Clearing of fractions, $\quad 969 + 38y - 15y = 900.$

Transposing and uniting, $\quad\quad\quad 23y = -69.$

$$y = -3. \quad \text{Ans.}$$

Substituting this value in equation (3), we have

$$x = \frac{51 - 6}{5} = 9. \quad \text{Ans.}$$

(231) (a) $2x^2 - 27x = 14.$

$$x^2 - \frac{27x}{2} = 7.$$

$$x^2 - \frac{27x}{2} + \left(\frac{27}{4}\right)^2 = 7 + \left(\frac{27}{4}\right)^2 = \frac{841}{16}.$$

$$x - \frac{27}{4} = \pm \frac{29}{4}.$$

$$x = \frac{56}{4} = 14,$$

$$\text{or} \quad x = -\frac{2}{4} = -\frac{1}{2}.$$

Hence, $x = 14, \text{ or } -\dfrac{1}{2}.$ Ans

(b) $x^2 - \dfrac{2x}{3} + \dfrac{1}{12} = 0.$

Transposing, $x^2 - \dfrac{2x}{3} = -\dfrac{1}{12}.$

Completing the square,

$$x^2 - \frac{2x}{3} + \left(\frac{1}{3}\right)^2 = -\frac{1}{12} + \frac{1}{9} = \frac{1}{36}.$$

Extracting the square root,

$$x - \frac{1}{3} = \pm \frac{1}{6}.$$

Transposing, $x = \dfrac{1}{3} + \dfrac{1}{6} = \dfrac{1}{2},$

$$\text{or} \quad x = \frac{1}{3} - \frac{1}{6} = \frac{1}{6}.$$

Therefore, $x = \dfrac{1}{2} \text{ or } \dfrac{1}{6}.$ Ans.

(c) $x^2 + ax = bx + ab.$

Transposing and factoring,

$$x^2 + (a - b)x = ab.$$

$$x^2 + (a - b)x + \left(\frac{a - b}{2}\right)^2 = ab + \left(\frac{a - b}{2}\right)^2 =$$

$$\frac{4ab + a^2 - 2ab + b^2}{4} = \frac{a^2 + 2ab + b^2}{4}.$$

Extracting square root,

$$x + \frac{a-b}{2} = \pm \frac{a+b}{2}.$$

$$x = -\frac{a-b}{2} + \frac{a+b}{2} = b,$$

or $$x = -\frac{a-b}{2} - \frac{a+b}{2} = -a.$$

Therefore, $x = b$, or $-a$.   Ans.

(232)   Let x = rate of current.
y = rate of rowing.

Down stream, the rowers are aided by the current, so $x + y = 12$.

Since it takes them twice as long to row a given distance *up* stream as it does down stream, they will go only $\frac{1}{2}$ as far in 1 hour, or $\frac{1}{2}$ of $12 = 6$ miles per hour up stream.

$$x + y = 12. \quad (1)$$
$$-x + y = 6. \quad (2)$$

Subtracting, $2x \quad\quad = 6$, and $x = 3$ miles per hour.
Ans.

(233)   (a) $\dfrac{10x+3}{3} - \dfrac{6x-7}{2} = 10(x-1).$

Reducing the last member to a simpler form, the equation becomes

$$\frac{10x+3}{3} - \frac{6x-7}{2} = 10x - 10.$$

Clearing of fractions by multiplying each term of both members by 6, the L. C. M. of the denominators, and changing the sign of each term of the numerator of the second fraction, since it is preceded by the minus sign (Art. **567**), we have

$$20x + 6 - 18x + 21 = 60x - 60.$$

Transposing terms, $20x - 18x - 60x = -60 - 21 - 6$.
Combining like terms, $-58x = -87$.
Changing signs, $58x = 87$;

whence, $x = \dfrac{87}{58} = 1\dfrac{1}{2}.$ Ans.

(*b*) $(a^2 + x)^2 = x^2 = 4a^2 + a^4.$

Performing the operation indicated in the first member, the equation becomes

$$a^4 + 2a^2 x + x^2 = x^2 + 4a^2 + a^4.$$

Canceling x^2 (Art. **562**) and transposing,

$$2a^2 x = 4a^2 + a^4 - a^4.$$

Combining like terms, $2a^2 x = 4a^2.$

Dividing by $2a^2$, $x = 2.$ Ans.

(*c*) $\dfrac{x-1}{x-2} - \dfrac{x+1}{x+2} = \dfrac{3}{x^2 - 4}.$

Clearing of fractions, the equation becomes

$$(x - 1)(x + 2) - (x + 1)(x - 2) = 3.$$

Expanding, $x^2 + x - 2 - x^2 + x + 2 = 3.$

Uniting terms, $2x = 3.$

$$x = \frac{3}{2} = 1\frac{1}{2}.$$ Ans.

(234) $11x + 3y = 100.$ (1)
                    $4x - 7y = 4.$ (2)

Since the signs of the terms containing x in each equation are alike, x may be eliminated by subtraction. If the first equation be multiplied by 4, and the second by 11, the coefficients in each case will become equal. Hence,

Multiplying (1) by 4, $44x + 12y = 400.$ (3)
Multiplying (2) by 11, $44x - 77y = \ \ 44.$ (4)

Subtracting (4) from (3), $89y = 356.$
                    $y = 4.$ Ans.

Substituting this value for y in (2),

$$4x - 28 = 4.$$
$$4x = 32.$$
$$x = 8.$$ Ans.

(235) (*a*) $y^{\frac{5}{3}} = 243.$

Extracting fifth root of both terms,

$$y^{\frac{1}{3}} = 3.$$

Cubing both terms, $y = 3^3 = 27.$ Ans.

(b)
$$x^{10} + 31x^5 - 10 = 22,$$
$$\text{or} \quad x^{10} + 31x^5 = 32.$$

Completing the square,

$$x^{10} + 31x^5 + \left(\frac{31}{2}\right)^2 = 32 + \left(\frac{31}{2}\right)^2.$$

$$x^{10} + 31x^5 + \left(\frac{31}{2}\right)^2 = 32 + \frac{961}{4} = \frac{1,089}{4}.$$

Extracting square root, $\quad x^5 + \frac{31}{2} = \pm \frac{33}{2}.$

Transposing, $\quad x^5 = -\frac{31}{2} + \frac{33}{2} = \frac{2}{2},$ or 1.

$$x^5 = -\frac{31}{2} - \frac{33}{2} = -\frac{64}{2} = -32;$$

whence, $\quad x = \sqrt[5]{1} = 1,$

or $\quad x = \sqrt[5]{-32} = -2.$ Ans. (Art. **600.**)

(c) $\quad x^2 - 4x^{\frac{3}{2}} = 96.$

Completing the square,
$$x^2 - 4x^{\frac{3}{2}} + 4 = 96 + 4 = 100.$$

Extracting square root, $x^{\frac{3}{2}} - 2 = \pm 10.$

Transposing and combining, $x^{\frac{3}{2}} = 12,$ or $-8.$

But, $\qquad\qquad\qquad x^{\frac{3}{2}} = \sqrt{x^3} = 12,$ or $-8.$

Removing the radical, $\quad x^3 = 144,$ or $(-8)^2.$

$$\sqrt[3]{144} = \sqrt[3]{8} \times \sqrt[3]{18} = 2\sqrt[3]{18}. \quad \text{(Art. \textbf{538.})}$$

$$\sqrt[3]{(-8)^2} = (-8)^{\frac{2}{3}}.$$

Hence, $\qquad\qquad x = 2\sqrt[3]{18},$ or $(-8)^{\frac{2}{3}}.$ Ans.

(**236**) (a) The value of a^0 is the same as 1. (Arts. **438** and **439.**)

(b) $\dfrac{a^0}{a^{-1}} = a.$ Ans. (Art. **530.**)

(c) $\sqrt{(3x^2 + 5xy^2 + 6x^2y)^2} = 3x^2 + 5xy^2 + 6x^2y = 3 \times 2^2 + 5 \times 2 \times 4^2 + 6 \times 2^2 \times 4 = 12 + 640 + 96 = 748,$ when $x = 2,$ and $y = 4.$ Ans.

(237) *(a)* $\quad \dfrac{6x+1}{15} - \dfrac{2x-4}{7x-16} = \dfrac{2x-1}{5}.$

Clearing of fractions,

$$(6x+1)(7x-16) - 15(2x-4) = 3(2x-1)(7x-16)$$

Removing parentheses and expanding,

$$42x^2 - 89x - 16 - 30x + 60 = 42x^2 - 117x + 48.$$

Canceling $42x^2$ (Art. **562**) and transposing,

$$117x - 89x - 30x = 16 - 60 + 48$$

Combining terms, $\qquad\qquad -2x = 4.$

$$x = -2. \quad \text{Ans.}$$

(b) $\qquad\qquad \dfrac{ax^2}{c-bx} + a + \dfrac{ax}{b} = 0.$

$$abx^2 + abc - ab^2x + acx - abx^2 = 0.$$

Transposing and uniting, $acx - ab^2x = -abc.$

$$a(c - b^2)x = -abc.$$

$$x = -\dfrac{abc}{a(c-b^2)}.$$

Canceling the common factor a and changing two of the signs of the fraction (Art. **482**),

$$x = \dfrac{bc}{b^2 - c}. \quad \text{Ans.}$$

(c) $\qquad\qquad \dfrac{\sqrt{x}-3}{\sqrt{x}+7} = \dfrac{\sqrt{x}-4}{\sqrt{x}+1}.$

Clearing of fractions,

$$(\sqrt{x}-3)(\sqrt{x}+1) = (\sqrt{x}+7)(\sqrt{x}-4) =$$

$$x - 2\sqrt{x} - 3 = x + 3\sqrt{x} - 28.$$

Transposing and canceling x (Art. **562**),

$$-2\sqrt{x} - 3\sqrt{x} = 3 - 28.$$

$$-5\sqrt{x} = -25.$$

$$\sqrt{x} = 5.$$

$$x = 25. \quad \text{Ans.}$$

(238) (a) $\sqrt{\frac{3}{2}}$ by Art. **540** $= \sqrt{\frac{3}{2} \times \frac{2}{2}} = \sqrt{\frac{6}{4}} = \frac{1}{2}\sqrt{6}.$

Ans.

(b) $\frac{3}{11}\sqrt{\frac{4}{7}} = \frac{3}{11}\sqrt{\frac{4}{7} \times \frac{7}{7}} = \frac{3}{11} \times \frac{2}{7}\sqrt{7} = \frac{6}{77}\sqrt{7}.$ Ans.

(c) $z\sqrt[3]{\frac{2x}{z}} = z\sqrt[3]{\frac{2x}{z} \times \frac{z^2}{z^2}} = \frac{z}{z}\sqrt[3]{2xz^2} = \sqrt[3]{2xz^2}.$ Ans.

(239) (a) $\quad \dfrac{9x + 20}{36} = \dfrac{4(x-3)}{5x-4} + \dfrac{x}{4} = ?$

When the denominators contain both simple and compound expressions, it is best to remove the simple expressions first, and then remove each compound expression in order. Then, after each multiplication, the result should be reduced to the simplest form.

Multiplying both sides by 36,

$$9x + 20 = \frac{144(x-3)}{5x-4} + 9x,$$

or $\quad \dfrac{144x - 432}{5x - 4} = 20.$

Clearing of fractions,

$$144x - 432 = 100x - 80.$$

Transposing and combining,

$$44x = 352;$$

whence, $\quad x = 8.$ Ans.

(b) $\quad ax - \dfrac{3a - bx}{2} = \dfrac{1}{2}$ becomes, when cleared of fractions,

$$2ax - 3a + bx = 1.$$

Transposing and uniting terms,

$$2ax + bx = 3a + 1.$$

Factoring, $\qquad (2a + b)x = 3a + 1;$

whence, $\qquad x = \dfrac{3a + 1}{2a + b}.$ Ans.

(c) $am - b - \dfrac{ax}{b} + \dfrac{x}{m} = 0$, when cleared of fractions $=$

$$abm^2 - b^2m - amx + bx = 0.$$

Transposing, $bx - amx = b^2m - abm^2.$

Factoring, $(b - am)x = bm(b - am);$

whence, $x = \dfrac{bm(b - am)}{(b - am)} = bm.$ Ans.

(240) $x + y = 13.$ (1)

$xy = 36.$ (2)

Squaring (1) we have

$$x^2 + 2xy + y^2 = 169.$$ (3)

Multiplying (2) by 4, $4xy = 144.$ (4)

Subtracting (4) from (3),

$$x^2 - 2xy + y^2 = 25.$$ (5)

Extracting the square root of (5),

$$x - y = \pm 5.$$ (6)

Adding (6) and (1), $2x = 18$ or $8,$

$x = 9$ or $4.$ Ans.

Substituting the value of x in (1),

$$9 + y = 13,$$

or $4 + y = 13;$

whence, $y = 4,$ }

or $y = 9.$ }   Ans.

(241) $x^2 - y^2 = 98.$ (1)

$x - y = 2.$ (2)

From (2), $x = 2 + y.$ (3)

Substituting the value of x in (1),

$$8 + 12y + 6y^2 + y^3 - y^3 = 98.$$

Combining and transposing,

$$6y^2 + 12y = 90.$$

$$y^2 + 2y = 15.$$

$$y^2 + 2y + 1 = 15 + 1 = 16.$$

$$y + 1 = \pm 4.$$

$$y = 3, \text{ or } - 5.$$ Ans.

Substituting the value of y in (3), $x = 5$, or $- 3.$ Ans.

(**242**)   Let x = the whole quantity.

Then, $\dfrac{2x}{3} + 10$ = the quantity of niter.

$\dfrac{x}{6} - 4\dfrac{1}{2}$ = the quantity of sulphur.

$\dfrac{1}{7}\left(\dfrac{2x}{3} + 10\right) - 2$ = the quantity of charcoal.

Hence, $x = \dfrac{2x}{3} + 10 + \dfrac{x}{6} - 4\dfrac{1}{2} + \dfrac{1}{7}\left(\dfrac{2x}{3} + 10\right) - 2.$

Clearing of fractions and expanding terms,

$42x = 28x + 420 + 7x - 189 + 4x + 60 - 84.$

Transposing,

$42x - 28x - 7x - 4x = 420 - 189 + 60 - 84.$

$$3x = 207.$$
$$x = 69 \text{ lb.}\quad \text{Ans.}$$

(**243**)   Let x = number of revolutions of hind wheel.

Then, $51 + x$ = number of revolutions of fore wheel.

Since, in making these revolutions both wheels traveled the same distance, we have

$$16x = 14\,(51 + x).$$
$$16x = 714 + 14x.$$
$$2x = 714.$$
$$x = 357.$$

Since the hind wheel made 357 revolutions, and since the distance traveled for each revolution is equal to the circumference of the wheel, or 16 feet, the whole distance traveled = 357 × 16 ft. = 5,712 feet.   Ans.

(**244**)   (*a*)   Transposing,

$$5x^2 - 2x^2 = 24 + 9.$$

Uniting terms, $3x^2 = 33.$

$$x^2 = 11.$$

Extracting the square root of both members,

$$x = \pm \sqrt{11}.\quad \text{Ans.}$$

(b)
$$\frac{3}{4x^2} - \frac{1}{6x^2} = \frac{7}{3}.$$

Clearing of fractions, $9 - 2 = 28x^2.$

Transposing terms, $28x^2 = 7.$

$$x^2 = \frac{1}{4}.$$

Extracting the square root of both members,

$$x = \pm \frac{1}{2}. \quad \text{Ans.}$$

(c) $\quad \dfrac{x^2}{5} - \dfrac{x^2 - 10}{15} = 7 - \dfrac{50 + x^2}{25}.$

Clearing of fractions by multiplying each term of both members by 75, the L. C. M. of the denominators, and expanding,

$$15x^2 - 5x^2 + 50 = 525 - 150 - 3x^2.$$

Transposing and uniting terms,

$$13x^2 = 325.$$

Dividing by 13, $\qquad x^2 = \dfrac{325}{13} = 25,$

$$\text{or} \quad x = \pm 5. \quad \text{Ans.}$$

(245) $\qquad 4x + 3y = 48.$ $\qquad$ (1)

$$-3x + 5y = 22. \qquad (2)$$

From (1), $\qquad\qquad y = \dfrac{48 - 4x}{3}. \quad$ (3)

From (2), $\qquad\qquad y = \dfrac{22 + 3x}{5}. \quad$ (4)

Placing (3) and (4) equal to each other,

$$\frac{48 - 4x}{3} = \frac{22 + 3x}{5}.$$

Clearing of fractions,

$$240 - 20x = 66 + 9x.$$

Transposing and uniting terms,

$$-29x = -174,$$
$$\text{or} \qquad x = 6. \quad \text{Ans.}$$

Substituting this value in (4),

$$y = \frac{22 + 18}{5} = 8. \quad \text{Ans.}$$

(246) Let $x =$ speed of one.

$x + 10 =$ speed of other.

Then, $\dfrac{1,200}{x} =$ number of hours one train required.

$\dfrac{1,200}{x + 10} =$ number of hours other train required.

$$\frac{1,200}{x} = \frac{1,200}{x + 10} + 10.$$

$$1,200x + 12,000 = 1,200x + 10x^2 + 100x.$$

$$- 10x^2 - 100x = - 12,000.$$

$$10x^2 + 100x = 12,000.$$

$$x^2 + 10x = 1,200.$$

$$x^2 + 10x + 25 = 1,200 + 25 = 1,225.$$

$$x + 5 = \pm\, 35.$$

$\left. \begin{array}{l} x = 30 \text{ miles per hour.} \\ x + 10 = 40 \text{ miles per hour.} \end{array} \right\}$ Ans.

(247) $\left. \begin{array}{l} 2x - \dfrac{y - 3}{5} - 4 = 0 \\[2mm] 3y + \dfrac{x - 2}{3} - 9 = 0 \end{array} \right\}$ cleared of fractions, becomes

$$10x - y + 3 - 20 = 0. \qquad (1)$$

$$9y + x - 2 - 27 = 0. \qquad (2)$$

Transposing and uniting, $10x - y = 17.$ $\qquad (3)$

$$x + 9y = 29. \qquad (4)$$

Multiplying (4) by 10 and subtracting (3) from the result,

$$\begin{array}{r} 10x + 90y = 290 \\ 10x - y = 17 \\ \hline 91y = 273 \\ y = 3. \quad \text{Ans.} \end{array}$$

Substituting value of y in (4),

$$x + 27 = 29.$$

$$x = 2. \quad \text{Ans.}$$

(248) (a) $\sqrt[3]{2} \times \sqrt[4]{3} = 2^{\frac{1}{3}} \times 3^{\frac{1}{4}}.$ (Art. **547.**)

$2^{\frac{4}{12}} \times 3^{\frac{3}{12}} = \sqrt[12]{2^4} \times \sqrt[12]{3^3} = \sqrt[12]{32 \times 27} = \sqrt[12]{864}.$ Ans.

(b) $\sqrt[4]{2ax} \times \sqrt[3]{ax^2} = (2ax)^{\frac{1}{4}} \times (ax^2)^{\frac{1}{3}} = \sqrt[12]{8a^3x^3} \times \sqrt[12]{a^4x^8} =$
$\sqrt[12]{8a^7x^{11}}$. Ans.

(c) $2\sqrt{xy} \times 3\sqrt[3]{x^2y} = 2 \times 3(xy)^{\frac{1}{2}} \times (x^2y)^{\frac{1}{3}} = 6\sqrt[6]{x^7y^5}$. Ans.

(**249**) Let $x =$ the part of the work which they all can do in 1 day when working together.

Then, since $\frac{1}{7\frac{1}{2}} = \frac{2}{15}$, $\frac{1}{5} + \frac{1}{6} + \frac{2}{15} = x$;

or, clearing of fractions and adding,

$$15 = 30x, \text{ and } x = \frac{1}{2}.$$

Since they can do $\frac{1}{2}$ the work in 1 day, they can do all of the work in 2 days.   Ans.

(**250**)   Let $x =$ value of first horse.
          $y =$ value of second horse.

If the saddle be put on the first horse, its value will be $x + 10$. This value is double that of the second horse, or $2y$, whence the equation, $x + 10 = 2y$.

If the saddle be put on the second horse, its value is $y + 10$. This value is \$13 less than the first, or $x - 13$, whence the equation, $y + 10 = x - 13$.

$$x + 10 = 2y. \qquad (1)$$
$$y + 10 = x - 13. \quad (2)$$

Transposing, $x - 2y = -10.$ (3)
          $-x + y = -23.$ (4)

Adding (3) and (4), $-y = -33.$

       $y = \$33$, or value of second horse. Ans.

Substituting in (1), $x + 10 = 66$;

       or $x = \$56$, or value of first horse. Ans.

(**251**)   Let $x =$ A's money.
          $y =$ B's money.

If A should give B \$5, A would have $x - 5$, and B, $y + 5$. B would then have \$6 more than A, whence the equation,

$$y + 5 - (x - 5) = 6. \qquad (1)$$

But if A received $5 from B, A would have $x + 5$, and B, $y - 5$, and 3 times his money, or $3(x + 5)$, would be $20 more than 4 times B's, or $4(y - 5)$, whence the equation,

$$3(x + 5) - 4(y - 5) = 20. \qquad (2)$$

Expanding equations (1) and (2),

$$y + 5 - x + 5 = 6. \qquad (3)$$
$$3x + 15 - 4y + 20 = 20. \qquad (4)$$

Transposing and combining,

$$y - x = -4. \qquad (5)$$
$$-4y + 3x = -15. \qquad (6)$$

Multiplying (5) by 4, and adding to (6),

$$4y - 4x = -16.$$
$$-4y + 3x = -15.$$
$$\overline{\qquad -x = -31.}$$
$$x = 31.$$

Substituting value of x in (5),

$$y - 31 = -4.$$
$$y = 27.$$

Hence, $\quad x = \$31,$ A's money. $\Big\}$ Ans.
$\qquad\quad y = \$27,$ B's money. $\Big\}$

(252) (*a*) $\qquad x^2 - 6x = 16.$

Completing the square (Art. **597**),

$$x^2 - 6x + 9 = 16 + 9.$$

Extracting the square root, $x - 3 = \pm 5.$

Transposing, $\qquad\qquad x = 8,$ or $-2.$ Ans.

(*b*) $\qquad\qquad x^2 - 7x = 8.$

$$x^2 - 7x + \left(\frac{7}{2}\right)^2 = 8 + \left(\frac{7}{2}\right)^2 = \frac{81}{4}.$$

$$x - \frac{7}{2} = \pm \frac{9}{2};$$

whence, $\qquad x = 8,$ or $-1.$ Ans.

(c) $9x^2 - 12x = 21.$

Dividing by 9, $x^2 - \dfrac{12x}{9} = \dfrac{21}{9},$

or $x^2 - \dfrac{4x}{3} = \dfrac{7}{3}.$

Completing the square,

$$x^2 - \frac{4x}{3} + \left(\frac{2}{3}\right)^2 = \frac{7}{3} + \frac{4}{9} = \frac{25}{9}.$$

Extracting square root, $x - \dfrac{2}{3} = \pm \dfrac{5}{3}.$

Transposing, $x = \dfrac{2}{3} + \dfrac{5}{3} = \dfrac{7}{3},$

or $x = \dfrac{2}{3} - \dfrac{5}{3} = -\dfrac{3}{3} = -1.$

Therefore, $x = \dfrac{7}{3}$, or $-1.$ Ans.

(253) $(c^{-\frac{1}{2}})^{-\frac{1}{2}} = c^{\frac{1}{4}}.$ Ans.   (Art. **526**, III.)

$(m\sqrt{n^3})^{-\frac{1}{2}} = m^{-\frac{1}{2}}(n^{\frac{3}{2}})^{-\frac{1}{2}} = m^{-\frac{1}{2}}n^{-\frac{3}{4}} = \dfrac{1}{m^{\frac{1}{2}}n^{\frac{3}{4}}}.$ Ans.

$(cd^{-2})^{\frac{1}{a}} = c^{\frac{1}{a}}d^{-\frac{2}{a}},$ or $\sqrt[a]{cd^{-2}},$ or $\sqrt[a]{\dfrac{c}{d^2}}.$ Ans.   (Art. **530**.)

(254)

Let $x =$ number of quarts of 90-cent wine in the mixture.

$y =$ number of quarts of 50-cent wine in the mixture

Then, $x + y = 60,$ (1)

and $90x + 50y = 4,500 = 75 \times 60.$ (2)

Multiplying (1) by 50,

$50x + 50y = 3,000.$ (3)

Subtracting (3) from (2),

$40x = 1,500;$

whence, $x = 37\frac{1}{2}$ qt.   Ans.

Multiplying (1) by 90, $90x + 90y = 5,400$ (4)

Subtracting (2), $\dfrac{90x + 50y = 4,500}{40y = \quad 900;}$ (2)

whence, $y = 22\frac{1}{2}$ qt.   Ans.

(255) Let $x =$ the numerator of the fraction.

$y =$ the denominator of the fraction.

Then, $\dfrac{x}{y} =$ the fraction.

From the conditions, $\qquad \dfrac{2x}{y+7} = \dfrac{2}{3}$, $\qquad$ (1)

and $\qquad \dfrac{x+2}{2y} = \dfrac{3}{5}$. $\qquad$ (2)

Clearing (1) and (2) of fractions, and transposing,

$$6x = 2y + 14, \quad (3)$$

and $\quad 5x = 6y - 10. \quad (4)$

Solving for x, $\qquad x = \dfrac{2y+14}{6} = \dfrac{y+7}{3}. \quad (5)$

$$x = \dfrac{6y-10}{5}. \qquad (6)$$

Equating (5) and (6), $\qquad \dfrac{y+7}{3} = \dfrac{6y-10}{5}.$

Clearing of fractions, $\quad 5y + 35 = 18y - 30$

whence, $\quad 13y = 65$,

or, $\qquad y = 5.$

Substituting this value of y in (3),

$$6x = 10 + 14 = 24;$$

whence, $x = 4.$

Therefore, the fraction is $\dfrac{4}{5}$. Ans.

(256) $\qquad$ Let $x =$ digit in tens place.

$y =$ digit in units place.

Then $10x + y =$ the number.

From the conditions of the example,

$$10x + y = 4(x+y) = 4x + 4y;$$

whence, $\quad 3y = 6x$,

or $\qquad y = 2x.$

From the conditions of the example,

$$10x + y + 18 = 10y + x;$$

whence, $\quad 9y - 9x = 18.$

Substituting the value of y, found above,

$$18x - 9x = 18;$$

whence, $x = 2.$

$$y = 2x = 4.$$

Hence, the number $= 10x + y = 20 + 4 = 24$. Ans.

(257) Let $x =$ greater number.

$y =$ less number.

Then, $x + 4 = 3\frac{1}{4}y,$ (1)

and $y + 8 = \dfrac{x}{2}.$ (2)

Clearing of fractions, $4x + 16 = 13y,$

and $2y + 16 = x;$

whence, $13y - 4x = 16.$ (3)

$2y - x = -16.$ (4)

Multiplying (4) by 4, and subtracting from (3)

$$5y = 80,$$

or $y = 16$. Ans.

Substituting in (4), $32 - x = -16;$

whence, $x = 48$. Ans.

LOGARITHMS.

(258) First raise $\frac{200}{100}$ to the .29078 power. Since $\frac{200}{100} = 2$,

$\left(\frac{200}{100}\right)^{.29078} = 2^{.29078}$, and $\log 2^{.29078} = .29078 \times \log 2 = .29078 \times$

$.30103 = .08753$. Number corresponding $= 1.2233$. Then,

$$1 - \left(\frac{200}{100}\right)^{.29078} = 1 - 1.2233 = -.2233.$$

We now find the product required by adding the logarithms of 351.36, 100, 24, and .2233, paying no attention to the negative sign of .2233 until the product is found. (Art. **647.**)

$$\text{Log } 351.36 = 2.54575$$
$$\log \quad 100 = 2$$
$$\log \quad\;\; 24 = 1.38021$$
$$\log \;\; .2233 = \overline{1}.34889$$
$$sum = \overline{5.27485} =$$

$$\log 351.36 \times 100 \times 24 \left(1 - \left(\frac{200}{100}\right)^{.29078}\right)$$

Number corresponding $= 188,300$.

The number is negative, since multiplying positive and negative signs gives negative; and the sign of .2233 is minus. Hence,

$$x = -188,300. \quad \text{Ans.}$$

(259) (*a*) Log $2,376 = 3.37585$. Ans. (See Arts. **625** and **627.**)

(*b*) Log $.6413 = \overline{1}.80706$. Ans.

(*c*) Log $.0002507 = \overline{4}.39915$. Ans.

(260) (*a*) Apply rule, Art. **652.**

$$\text{Log}\ \ 755.4 = 2.87818$$
$$\log .00324 = \bar{3}.51055$$
$$\textit{difference} = 5.36763 = \text{logarithm of quotient.}$$

The mantissa is not found in the table. The next less mantissa is 36754. The difference between this and the next greater mantissa is $773 - 754 = 19$, and the P. P. is $763 - 754 = 9$. Looking in the P. P. section for the column headed 19, we find opposite 9.5, 5, the fifth figure of the number. The fourth figure is 1, and the first three figures 233; hence, the figures of the number are 23315. Since the characteristic is 5, $755.4 \div .00324 = 233,150.$ Ans.

(*b*) Apply rule, Art. **652.**

$$\text{Log}\ \ \ .05555 = \bar{2}.74468$$
$$\log .0008601 = \bar{4}.93455$$
$$\textit{difference} = 1.81013 = \text{logarithm of quotient.}$$

The number whose logarithm is 1.81013 equals 64.584.

Hence, $.05555 \div .0008601 = 64.584.$ Ans.

(*c*) Apply rule, Art. **652.**

$$\text{Log}\ \ 4.62 = \ \ .66464$$
$$\log .6448 = \bar{1}.80943$$
$$\textit{difference} = \ \ .85521 = \text{logarithm of quotient.}$$

Number whose logarithm $= .85521 = 7.1648.$

Hence, $4.62 \div .6448 = 7.1648.$ Ans.

(261) $x^{.74} = \dfrac{238 \times 1,000}{.0042^{.0002}}.$

$$\text{Log}\ \ \ 238 = 2.37658$$
$$\log 1,000 = 3.$$
$$\textit{sum} = 5.37658 = \log (238 \times 1,000).$$

$$\text{Log} .0042 = \bar{3}.6\,2\,3\,2\,5$$
$$\underline{\phantom{\text{Log} .0042 = \bar{3}}.6\,6\,0\,2}$$
$$1\,2\,4\,6\,5\,0$$
$$3\,7\,3\,9\,5\,0$$
$$\underline{3\,7\,3\,9\,5\,0}$$
$$.4\,1\,1\,4\,6\,9\,6\,5\,0 \text{ or } .41147.$$

$$.6602$$
$$-3$$
$$-1.9806 = \text{characteristic.}$$

Adding, $.41147$
$$-1.9806$$
$$\overline{2}.43087 \quad (\text{See Art. } \mathbf{659.})$$

Then, $\log\left(\dfrac{238 \times 1,000}{.0042^{.0002}}\right) = 5.37658 - \overline{2}.43087 = 6.94571 =$

$74 \log x$; whence, $\log x = \dfrac{6.94571}{.74} = 9.38609.$ Number

whose logarithm $= 9.38609$ is $2,432,700,000 = x.$ Ans.

(262) Log $.00743 = \overline{3}.87099.$
log $.006 \quad = \overline{3}.77815.$

$\sqrt[5]{.00743} = \log .00743 \div 5$ (Art. **662**), and $\sqrt[.6]{.006} = \log$
$.006 \div .6$. Since these numbers are wholly decimal, we
apply Art. **663.**

$$5\,)\,\overline{3}.87099$$
$$\overline{1.57419} = \log \sqrt[5]{.00743}.$$

The characteristic $\overline{3}$ will not contain 5. We then add $\overline{2}$
to it, making $\overline{5}$. 5 is contained in $\overline{5}$, $\overline{1}$ times. Hence, the
characteristic is $\overline{1}$. Adding the same number, 2, to the
mantissa, we have 2.87099. 2.87099 $\div 5 = .57419$. Hence,
$\log \sqrt[5]{.00743} = \overline{1}.57419.$

$.6\,)\,\overline{3}.77815$ .6 is contained in $\overline{3}$, -5 times.

$\overline{5}.$ .6 is contained in .77815, 1.29691 times.

1.29691

$sum = -\overline{4}.29691 = \sqrt[.6]{.006}.$

Log $\sqrt[5]{.00743} = \overline{1}.57419$
log $\sqrt[.6]{.006} \quad = \overline{4}.29691$

difference $= 3.27728 = $ log of quotient.

Number corresponding $= 1,893.6.$

Hence, $\sqrt[5]{.00743} \div \sqrt[.6]{.006} = 1,893.6.$ Ans.

(263)   Apply rule, Art. **647.**

$$\text{Log} \quad 1{,}728 = 3.23754$$
$$\log\ .00024 = \overline{4}.38021$$
$$\log\ \ .7462 = \overline{1}.87286$$
$$\log\ \ 302.1 = 2.48015$$
$$\log\ 7.6094 = \underline{\ \ .88135}$$
$$sum = \overline{2}.85211 =$$

log $(1{,}728 \times .00024 \times .7462 \times 302.1 \times 7.6094)$.   Number whose logarithm is $2.85211 = 711.40$, the product.   Ans.

(264) $\text{Log}\ \sqrt{5.954} = \ .77481 \div 2 = .38741$
$$\log\ \sqrt[3]{61.19} = 1.78668 \div 3 = \underline{.59556}$$
$$sum = \qquad\qquad\qquad .98297$$

$$\log\ \sqrt[5]{298.54} = 2.47500 \div 5 = .49500.$$

Then, $\dfrac{\sqrt{5.954} \times \sqrt[3]{61.19}}{\sqrt[5]{298.54}} = \log\ (\sqrt{5.954} \times \sqrt[3]{61.19}) - \log$

$\sqrt[5]{298.54} = .98297 - .49500 = .48797 =$ logarithm of the required result.

Number corresponding $= 3.0759$.   Ans.

(265) $\sqrt[7]{.0532864} = \log\ .0532864 \div 7.$
Log $.0532864 = \overline{2}.72661.$
Adding $\overline{5}$ to characteristic $\overline{2} = \overline{7}.$
Adding 5 to mantissa $= 5.72662.$
$\overline{7} \div 7 = \overline{1}.$
$5.72661 \div 7 = .81809$, nearly.
Hence, log $\sqrt[7]{.0532864} = \overline{1}.81809.$
Number corresponding to log $\overline{1}.81809 = .65780.$ Ans.

(266)   (a) $32^{4.8}$. 1.50515
Log $32 = 1.50515.$ $\underline{4.8}$
$$1204120$$
$$602060$$
$$\overline{7.224720}$$

7.22472 is the logarithm of the required power.   (Art. **657.**)

Number whose logarithm $= 7.22472$ is $16{,}777{,}000.$

Hence, $32^{4.8} = 16{,}777{,}000.$ Ans.

(b) .76^{'.«}.

Log .76 = $\overline{1}$.88081.

(See Arts. **658** and **659**.)

$$\overline{1} + .88081$$
$$3.62$$
$$\overline{176162}$$
$$528486$$
$$264243$$
$$\overline{3.1885322}$$
$$-3.62$$
$$\overline{\overline{1}.56853} = \log .37028.$$

Hence, .76^{'.«} $\rightleftharpoons$ .37028.   Ans.

(c) .84^{·»}.

Log .84 = $\overline{1}$.92428.

$$\overline{1} + .92428$$
$$.38$$
$$\overline{739424}$$
$$277284$$
$$\overline{.3512264}$$
$$-.38$$
$$\overline{\overline{1}.97123} = \log .93590.$$

Hence, .84^{·»} = .93590.   Ans.

(267)   Log $\sqrt[6]{\dfrac{1}{249}} - \log \sqrt[5]{\dfrac{23}{71}} = $ logarithm of answer.

Log $\sqrt[6]{\dfrac{1}{249}} = \dfrac{1}{6}(\log 1 - \log 249) = \dfrac{1}{6}(0 - 2.39620) = -.39937$

$= $ (adding $+ 1$ and $- 1$)$\overline{1}.60063$.

Log $\sqrt[5]{\dfrac{23}{71}} = \dfrac{1}{5}(\log 23 - \log 71) = \dfrac{1}{5}(1.36173 - 1.85126) =$

$\dfrac{1}{5}(-.48953) = -.097906 = $ (adding $+ 1$ and $- 1$) $\overline{1}.902094$,

or $\overline{1}.90209$ when using 5-place logarithms.

Hence, $\overline{1}.60063 - \overline{1}.90209 = \overline{1}.69854 = \log .49950$. Therefore,

$$\sqrt[6]{\dfrac{1}{249}} \div \sqrt[5]{\dfrac{23}{71}} = .49950.$$ Ans.

(268)   The mantissa is not found in the table.   The next less mantissa is .81291; the difference between this and the next greater mantissa is $298 - 291 = 7$, and the P. P. is $.81293 - .81291 = 2$. Looking in the P. P. section for the column headed 7, we find opposite 2.1, 3, the fifth figure of the number; the fourth figure is 0, and the first three figures, 650. Hence, the number whose logarithm is .81293 is 6.5003. Ans.

$2.52460 =$ logarithm of 334.65. Ans. (See Art. **640.**)

$\overline{1}.27631 =$ logarithm of .18893. Ans. We choose 3 for the fifth figure because, in the proportional parts column headed 23, 6.9 is nearer 8 than 9.2.

(**269**) The most expeditious way of solving this example is the following:

$$p\, v^{1\cdot41} = p_1 v_1{}^{1\cdot41}, \text{ or } v_1 = \sqrt[1\cdot41]{\frac{p\,v^{1\cdot41}}{p_1}} = v \sqrt[1\cdot41]{\frac{p}{p_1}}.$$

Substituting values given, $v_1 = 1.495 \sqrt[1\cdot41]{\dfrac{134.7}{16.421}}.$

Log $v_1 =$ log $1.495 + \dfrac{\log 134.7 - \log 16.421}{1.41} = .17464 +$

$\dfrac{2.12937 - 1.21540}{1.41} = .17464 + .64821 = .82285 =$ log $6.6504;$

whence, $v_1 = 6.6504.$ Ans.

(**270**) Log $\sqrt[5]{\dfrac{7.1895 \times 4{,}764.2^2 \times 0.00326^5}{.000489 \times 457^3 \times .576^2}} = \dfrac{1}{5}[\log 7.1895$

$+ 2 \log 4{,}764.2 + 5 \log .00326 - (\log .000489 + 3 \log 457 + 2$

$\log .576)] = \dfrac{\overline{5}.77878 - 4.18991}{5} = \overline{2}.31777 =$ log .020786. Ans

$$\begin{aligned}
\text{Log } 7.1895 = & \quad .85670 \\
2 \log 4{,}764.2 = 2 \times 3.67799 = & \quad 7.35598 \\
5 \log .00326 = 5 \times \overline{3}.51322 = & \quad \overline{13}.56610 \\
\hline
sum = & \quad \overline{5}.77878
\end{aligned}$$

$$\begin{aligned}
\text{Log } .000489 = & \quad \overline{4}.68931 \\
3 \log 457 = 3 \times 2.65992 = & \quad 7.97976 \\
2 \log .576 = 2 \times \overline{1}.76042 = & \quad \overline{1}.52084 \\
\hline
sum = & \quad 4.18991.
\end{aligned}$$

(**271**) Substituting the values given,

$$p = \frac{\frac{8,000}{960,000} \times \left(\frac{3}{16}\right)^{2.18}}{120 \times 2.25} = \frac{8,000\left(\frac{3}{16}\right)^{2.18}}{2.25}$$

Log $p = \log 8,000 + 2.18 \log \frac{3}{16} - \log 2.25 = 3.90309 + 2.18$

$(\log 3 - \log 16) - .35218 = 3.55091 + 2.18 \times (.47712 - 1.20412) = 1.96605 = \log 92.480.$ Ans.

(**272**) Solving for t, $t = \sqrt[2.18]{\frac{p\,l\,d}{960,000}}.$

Substituting values given,

$$t = \sqrt[2.18]{\frac{44}{\frac{160 \times 132 \times 2}{960,000}}} = \sqrt[2.18]{.044}.$$

$$\begin{array}{r} 8,000 \\ 3,000 \\ \hline 1,000 \end{array}$$

$$\text{Log } t = \frac{\log .044}{2.18} = \frac{\overline{2}.64345}{2.18} = \frac{-2.18 + .82345}{2.18} =$$

$\overline{1}.37773 = \log .23863.$ Ans.

GEOMETRY AND TRIGONOMETRY.

(**273**) When one straight line meets another straight line at a point between the ends, the sum of the two adjacent angles equals two right angles. Therefore, since one of the angles equals $\frac{4}{5}$ of a right angle, then, the other angle equals $\frac{10}{5}$, or two right angles, minus $\frac{4}{5}$. We have, then, $\frac{10}{5} - \frac{4}{5} = \frac{6}{5}$, or $1\frac{1}{5}$ right angles.

(**274**) The size of one angle is $\frac{1}{6}$ of two right angles, or $\frac{1}{3}$ of a right angle.

(**275**) The pitch being 4, the number of teeth in the wheel equals 4×12, or 48. The angle formed by drawing lines from the center to the middle points of two adjacent teeth equals $\frac{1}{48}$ of 4 right angles, or $\frac{1}{12}$ of a right angle.

(**276**) It is an isosceles triangle, since the sides opposite the equal angles are equal.

(**277**) An equilateral heptagon has seven equal sides; therefore, the length of the perimeter equals 7×3, or 21 inches.

(**278**) A regular decagon has 10 equal sides; therefore, the length of one side equals $\frac{40}{10}$, or 4 inches.

(**279**) The sum of all the interior angles of any polygon equals two right angles, multiplied by the number of sides in the polygon, less two. As a regular dodecagon has 12 equal sides, the sum of the interior angles equals two right angles × 10 (= 12 − 2), or 20 right angles. Since there are 12 equal angles, the size of any one of them equals 20 ÷ 12, or $1\frac{2}{3}$ right angles.

(**280**)   Equilateral triangle.

(**281**)   No, since the sum of the two smaller sides is not greater than the third side.

(**282**)   No, since the sum of the three smaller sides is not greater than the fourth side.

FIG. 1.

(**283**)   Since the two angles A and C, Fig. 1, are equal, the triangle is isosceles, and a line drawn from the vertex B will bisect the line $A\,C$, the length of which is 7 inches; therefore,

$$A\,D = D\,C = \frac{7}{2} = 3\frac{1}{2} \text{ in.}\quad \text{Ans.}$$

(**284**)   The length of the line $= \sqrt{12^2 - 9^2} + \sqrt{15^2 - 9^2}$, or 19.94 inches.

(**285**)   The sum of the three angles is equal to $\frac{8}{4}$, or 2 right angles; therefore, since the sum of two of them equals $\frac{5}{4}$ of a right angle, the third angle must equal $\frac{8}{4} - \frac{5}{4}$, or $\frac{3}{4}$ of a right angle.

(**286**)   One of the angles of an equiangular octagon is equal to $\frac{1}{8}$ of 12 right angles, or $1\frac{1}{2}$ right angles, since the sum of the interior angles of the equiangular octagon equals 12 right angles.

(**287**)   The sum of the acute angles of a right-angled triangle equals one right angle; therefore, if one of them equals $\frac{5}{8}$ of a right angle, the other equals $\frac{8}{8} - \frac{5}{8}$, or $\frac{3}{8}$ of a right angle.

(288) (See Art. **734.**)

(289) In Fig. 2, $AB = 4$ inches, and $AO = 6$ inches. We first find the length of DO. $DO = \sqrt{\overline{OA^2} - \overline{DA^2}}$; but $\overline{OA^2} = 6^2$, or 36, and $\overline{DA^2} = \left(\frac{4}{2}\right)^2$ or 4; therefore, $DO = \sqrt{36 - 4}$, or 5.657.

$DC = CO - DO$, or $DC = 6 - 5.657$, or .343 inch. In the right-angled triangle ADC, we have AC, which is the chord of one-half the arc ACB, equals $\sqrt{2^2 + .343^2}$, or 2.03 inches.

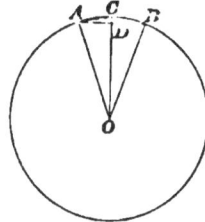
FIG. 2.

(290) The method of solving this is similar to the last problem.

$$DO = \sqrt{9-4}, \text{ or } 2.236. \quad DC = 3 - 2.236 = .764.$$
$$AC = \sqrt{2^2 + .764^2}, \text{ or } 2.14 \text{ inches.}$$

(291) Let HK of Fig. 3 be the section; then, $BI = 2$ inches, and $HK = 6$ inches, to find AB. $HI (= 3$ inches) being a mean proportional between the segments AI and IB, we have

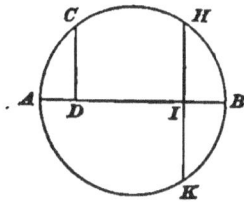
FIG. 3.

$$BI : HI :: HI : IA,$$
$$\text{or} \quad 2 : 3 :: 3 : IA.$$
$$\text{Therefore,} \quad IA = 4\frac{1}{2}.$$

$AB = AI + IB$; therefore, $AB = 4\frac{1}{2} + 2$, or $6\frac{1}{2}$ inches.

(292) Given $OC = 5\frac{3}{4}$ inches, and $OA = \frac{17}{2}$, or $8\frac{1}{2}$ inches, to find AB (see Fig. 4). CA, which is one-half the chord AB, equals

$$\sqrt{\overline{OA^2} - \overline{OC^2}};$$

therefore, $CA = \sqrt{(8\frac{1}{2})^2 - (5\frac{3}{4})^2}$, or 6.26 inches. Now, $AB = 2 \times CA$; therefore, $AB = 2 \times 6.26$, or 12.52 inches.

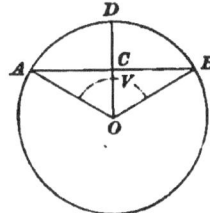
FIG. 4.

(293) The arc intercepted equals $\frac{3}{4}$ of 4, or 3 quadrants. As the inscribed angle is measured by one-half the intercepted arc, we have $\frac{3}{2} = 1\frac{1}{2}$ quadrants as the size of the angle.

(294) Four right angles $\div \frac{2}{7} = 4 \times \frac{7}{2}$, or 14 equal sectors.

(295) Since 24 inches equals the perimeter, we have $\frac{24}{8}$, or 3 inches, as the length of each side or chord.

Then, $2 \times \sqrt{\left(\frac{3}{2}\right)^2 + 3.62^2} = 7.84$ inches diameter.

(296) Given, $A\,C = \frac{A\,B}{2} = \frac{10.5}{2}$, or 5.25 inches. $A\,O$ and $A\,P = 13$ inches. (See Fig. 5.)

The required distance between the arcs $D\,D'$ is equal to $O\,A + A\,P - O\,P$. In the right-angled triangle $A\,C\,O$, we have

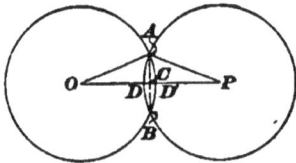

$$O\,C = \sqrt{\overline{A\,O^2} - \overline{A\,C^2}},$$

or $O\,C = \sqrt{169 - 27.5625} = 11.9$ inches.

FIG. 5.

Likewise, $C\,P = \sqrt{\overline{A\,P^2} - \overline{A\,C^2}} = 11.9$. $O\,P = O\,C + C\,P = 11.9 + 11.9 = 23.8$ inches. $O\,A + A\,P = 13 + 13 = 26$ inches. $26 - 23.8 = 2.2$ inches. Ans.

(297) Given $A\,P = 13$ inches, $O\,A = 8$ inches, and $A\,C = 5.25$ inches. Fig. 6.

$O\,C = \sqrt{\overline{A\,O^2} - \overline{A\,C^2}} = \sqrt{8^2 - 5.25^2} =$ 6.03 inches.

$C\,P = \sqrt{\overline{A\,P^2} - \overline{A\,C^2}} = 11.9$ inches.

$O\,P = O\,C + C\,P = 6.03 + 11.9 =$ 17.93 inches.

FIG. 6.

$D\,D' = O\,A + A\,P - O\,P = 8 + 13 - 17.93 = 3.07$ inches. Ans.

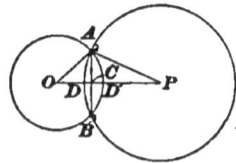

(298) $A\,B = 14$ inches, and $A\,E = 3\frac{1}{4}$ inches, Fig. 7. $C\,E = E\,D$ is a mean proportional between the segments

$A E$ and $E B$. Then,

$$A E : C E :: C E : EB,$$

or $\quad 3\frac{1}{4} : C E :: C E : 10\frac{3}{4},$

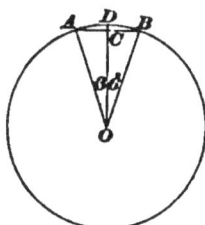

FIG. 7.

or $\quad \overline{C E}^{2} = 3\frac{1}{4} \times 10\frac{3}{4} = 34.9375.$

Extracting the square root, we have

$$C E = 5.91.$$

$2 \times C E = C D = 2 \times 5.91,$ or 11.82 inches. Ans.

(299) In 19° 19′ 19″ there are 69,559 seconds, and in 360°, or a circle, there are 1,296,000 seconds. Therefore, 69,559 seconds equal $\dfrac{69,559}{1,296,000}$, or .053672 part of a circle. Ans.

(300) In an angle measuring 19° 19′ 19″ there are 69,559 seconds, and in a quadrant, which is $\frac{1}{4}$ of 360°, or 90°, there are 324,000 seconds. Therefore, 69,559 seconds equal $\dfrac{69,559}{324,000}$, or .214688 part of a quadrant. Ans.

(301) Given, $O B = O A = \dfrac{23}{2}$, or $11\frac{1}{2}$ inches, and angle $A O B = \dfrac{1}{10}$ of 360°, or 36°. (See Fig. 8.) In the right-angled triangle $C O B$, we have

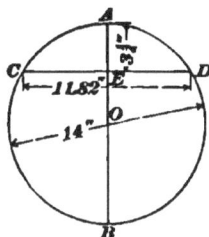

FIG. 8.

$\sin C O B = \dfrac{C B}{O B}$, or $C B = O B \times \sin C O B.$

Substituting the values of $O B$ and $\sin C O B$, we have

$$C B = 11\frac{1}{2} \times \sin 18°,$$

or $\quad C B = 11\frac{1}{2} \times .30902 = 3.55.$

Since $A B = 2 C B$, $A B = 2 \times 3.55 = 7.1$ inches.

The perimeter then equals $10 \times 7.1 = 71$ inches, nearly. Ans.

(302) $90° = 89°$ $59'$ $60''$

               $35°$ $24'$ $25.8''$
               _____
               $54°$ $35'$ $34.2''$ Ans.

(303) The side $BC = \sqrt{AB^2 - AC^2}$, or $BC = \sqrt{17.69^2 - 9.75^2} = \sqrt{217.8736} = 14$ ft. 9 in. To find the angle BAC, we have $\cos BAC = \dfrac{AC}{AB}$, or $\cos BAC = \dfrac{9.75}{17.69} = .55115$.

          $.55115$ equals the cos of $56° 33' 15''$.

Angle $ABC = 90° -$ angle BAC, or $90° - 56° 33' 15'' = 33° 26' 45''$.

(304) $159°$ $27'$ $34.6''$
               $25°$ $16'$ $8.7''$
               $3°$ $48'$ 53 '
               _____
               $188°$ $32'$ $36.3''$

(305)   Sin $17° 28' = .30015$.
               Sin $17° 27' = .29987$.

$.30015 - .29987 = .00028$, the difference for $1'$.

$.00028 \times \dfrac{37}{60} = .00017$, difference for $37''$.

$.29987 + .00017 = .30004 = \sin 17° 27' 37''$.

          Cos $17° 27' = .95398$.
          Cos $17° 28' = .95389$.

$.95398 - .95389 = .00009$, difference for $1'$.

$.00009 \times \dfrac{37}{60} = .00006$, difference for $37''$.

$.95398 - .00006 = .95392 = \cos 17° 27' 37''$.

          Tan $17° 28' = .31466$.
          Tan $17° 27' = .31434$.

$.31466 - .31434 = .00032$, difference for $1'$.

$.00032 \times \dfrac{37}{60} = .00020$, difference for $37''$.

$.31434 + .0002 = .31454 = \tan 17° 27' 37''$.

          Sin $17° 27' 37'' = .30004$ ⎫
          Cos $17° 27' 37'' = .95392$ ⎬ Ans.
          Tan $17° 27' 37'' = .31454$ ⎭

(306) From the vertex B, draw BD perpendicular to AC, forming the right-angled triangles ADB and BDC. In the right-angled triangle ADB, AB is known, and also the angle A. Hence, $BD = 26.583 \times \sin 36° \ 20' \ 43'' = 26.583 \times .59265 = 15.754$ feet. $AD = 26.583 \times \cos 36° 20' 43'' = 26.583 \times .80546 = 21.411$. $AC - AD = 40 - 21.411 = 18.589$ feet $= DC$. In the right-angled triangle BDC, the two sides BD and DC are known; hence, $\tan C = \dfrac{BD}{DC} = \dfrac{15.754}{18.589} = .84750$, and angle $C = 40° \ 16' \ 53''$. Ans.

$BC = \dfrac{BD}{\sin C} = \dfrac{15.754}{\sin 40° \ 16' \ 53''} = \dfrac{15.754}{.64654} = 24.37$, or 24 ft. 4.4 in. Ans.

Angle $B = 180° - (36° \ 20' \ 43'' + 40° \ 16' \ 53'') = 180° - 76° \ 37' \ 36'' = 103° \ 22' \ 24''$. Ans.

(307) This problem is solved exactly like problem No. 305.

$$\text{Sin of } 63° \ 4' \ 51.8'' = .89165.$$
$$\text{Cos of } 63° \ 4' \ 51.8'' = .45274.$$
$$\text{Tan of } 63° \ 4' \ 51.8'' = 1.96949.$$

(308)
$$.27038 = \sin \ 15° \ 41' \ 12.9''.$$
$$.27038 = \cos \ 74° \ 18' \ 47.1''.$$
$$2.27038 = \tan \ 66° \ 13' \ 43.2''.$$

(309) The angle formed by drawing radii to the extremities of one of the sides equals $\dfrac{360°}{11}$, or $32° \ 43' \ 38.2''$. Ans. The length of one side of the undecagon equals $\dfrac{4 \text{ ft. } 3 \text{ in.}}{11}$, or 4.6364 inches. The radius of the circle equals $\dfrac{\frac{1}{2} \text{ of } 4.6364}{\sin \text{ of } \frac{1}{2}(32° \ 43' \ 38.2'')} = \dfrac{2.3182}{.28173} = 8.23$ inches. Ans.

(310) If one of the angles is twice the given one, then it must be $2 \times (47° \ 13' \ 29'')$, or $94° \ 26' \ 58''$. Since there are two right angles, or 180°, in the three angles of a triangle, the third angle must be $180 - (47° \ 13' \ 29'' + 94° \ 26' \ 58'')$, or $38° \ 19' \ 33''$.

(**311**) If one of the angles is one-half as large as the given angle, then it must be $\frac{1}{2}$ of 75° 48′ 17″, or 37° 54′ 8.5″. The third angle equals 180° − (75° 48′ 17″ + 37° 54 8.5′), or 66° 17′ 34.5″.

(**312**) From the vertex B, draw BD perpendicular to AC, forming the two right-angled triangles ADB and BDC. In the right-angled triangle ADB, AB is known, and also the angle A. Hence, $BD = \sin A \times AB = \sin 54° 54′ 54″ \times 16\frac{5}{12} = .81830 \times 16\frac{5}{12} = 13.434$ feet.

Sine of angle $C = \dfrac{BD}{BC} = \dfrac{13.434}{13.542} = .99202$, and, hence, angle $C = 82° 45′ 30″$. Ans.

Angle $B = 180° − (54° 54′ 54″ + 82° 45′ 30″) = 180 − 137° 40′ 24″ = 42° 19′ 36″$. Ans.

$AD = AB \times \cos A = 16\frac{5}{12} \times \cos 54° 54′ 54″ = 16\frac{5}{12} \times .57479 = 9.43613$ ft.

$CD = BC \times \cos C = BC \times \cos 82° 45′ 30″ = 13\frac{13}{24} \times .12605 = 1.70692$ ft.

$AC = AD + CD = 9.43613 + 1.70692 = 11.143 = 11$ ft. $1\frac{3}{4}$ in. Ans.

(**313**) If one-third of a certain angle equals 14° 47′ 10″, then the angle must be $3 \times 14° 47′ 10″$, or 44° 21′ 30″. $2\frac{1}{2} \times 44° 21′ 30″$, or 110° 53′ 45″, equals one of the other two angles. The third angle equals 180° − (110° 53′ 45″ + 44° 21′ 30″), or 24° 44′ 45″.

(**314**) Given, $BC = 437$ feet and $AC = 792$ feet, to find the hypotenuse AB and the angles A and B.

$AB = \sqrt{\overline{AC}^2 + \overline{BC}^2} = \sqrt{792^2 + 437^2} = \sqrt{818,233} = 904$ ft. $6\frac{3}{4}$ in. Ans.

Tan $A = \dfrac{437}{792} = .55176$; therefore, $A = 28° \, 53' \, 18''$.   Ans.

Angle $B = 90° - 28° \, 53' \, 18''$, or $61° \, 6' \, 42''$.   Ans.

(315) In Fig. 9, angle $A\,O\,B = \dfrac{1}{8}$ of $360°$, or $45°$. Angle

$m\,O\,B = \dfrac{1}{2}$ of $45°$, or $22\dfrac{1}{2}°$. Side $A\,B =$

$\dfrac{1}{8}$ of 56 feet, or 7 feet. Now, in the

triangle $m\,O\,B$, we have the angle $m\,O\,B$

$= 22\dfrac{1}{2}°$, and $m\,B = \dfrac{7}{2}$, or $3\dfrac{1}{2}$ feet, given,

to find $O\,B$ and the angle $m\,B\,O$.

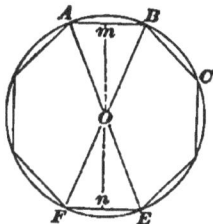

FIG. 9.

Sin $m\,O\,B = \dfrac{m\,B}{O\,B}$, or $O\,B = \dfrac{m\,B}{\sin m\,O\,B}$.

Substituting their values, $O\,B = \dfrac{3.5}{\sin 22\frac{1}{2}°} = \dfrac{3.5}{.38268} = 9.146$

feet.

$B\,F$, the diameter of the circle, equals $2 \times B\,O$; therefore,

$B\,F = 2 \times 9.146 = 18.292$ feet $= 18$ feet $3\dfrac{1}{2}$ inches.

Angle $B\,O\,m = 22° \, 30'$.

$B\,O\,m + O\,B\,m = 90°$.

Therefore, $O\,B\,m = 90° - B\,O\,m = 90° - 22° \, 30'$.

$= 67° \, 30'$.

$A\,B\,C = 2\,O\,B\,m - 2\,(67° \, 30') = 135°$.

Ans.

By Art. 703, the sum of the interior angles of an octagon

is $2\,(8 - 2) = 12$ right angles. Since the octagon is regular,

the interior angles are equal, and since there are eight of

them, each one is $\dfrac{12}{8} = 1\dfrac{1}{2}$ right angles. $1\dfrac{1}{2} \times 90° = 135°$.

(316) Lay off with a protractor the angle $A\,O\,C$ equal

to $67° \, 8' \, 49''$, Fig. 10. Tangent to the circle at A, draw the

line $A\,T$. Through the point C, draw the line $O\,C$, and

continue it until it intersects the line $A\,T$ at T. From C

draw the lines $C\,D$ and $C\,B$ perpendicular, respectively, to the radii $O\,E$ and $O\,A$. $C\,B$ is the sine, $C\,D$ the cosine, and $A\,T$ the tangent.

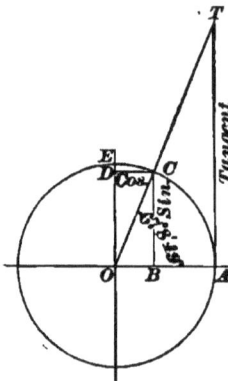

(**317**) Suppose that in Fig. 10, the line $A\,T$ has been drawn equal to 3 times the radius $O\,A$. From T draw $T\,O$; then, the tangent of $T\,O\,A$ $=\dfrac{T\,A}{O\,A}=3$. Where $T\,O$ cuts the circle at C, draw $C\,D$ and $C.B$ perpendicular, respectively, to $O\,E$ and $O\,A$. $C\,D$ is the cosine and $C\,B$ the sine.

FIG. 10. The angle corresponding to tan 3 is found by the table to equal $71°\ 33'\ 54''$; therefore, sin $71°\ 33'\ 54'' = .94868$ and cos $71°\ 33'\ 54'' = .31623$.

(**318**) The angle whose cos is .39278 $= 66°\ 52'\ 20'$.

Sin of $66°\ 52'\ 20'' =\ .91963$.

Tan of $66°\ 52'\ 20'' =\ 2.34132$.

For a circle with a diameter $4\frac{3}{4}$ times as large, the values of the above cos, sin, and tan will be

$$4\tfrac{3}{4} \times\ .39278 =\ 1.86570 \text{ cos.}$$

$$4\tfrac{3}{4} \times\ .91963 =\ 4.36824 \text{ sin.} \Big\} \text{ Ans.}$$

$$4\tfrac{3}{4} \times 2.34132 = 11.12127 \text{ tan.}$$

(**319**) See Fig. 11. Angle B $= 180° - (29°\ 21' + 76°\ 44'\ 18'') = 180° - 106°\ 5'\ 18'' = 73°\ 54'\ 42''$.

From C, draw $C\,D$ perpendicular to $A\,B$.

$A\,D = A\,C$ cos $A = 31.833$ $\times$ cos $29°\ 21' = 31.833 \times .87164$ $= 27.747$ ft. $C\,D = A\,C$ sin A $= 31.833 \times$ sin $29°\ 21' = 31.833$ $\times .49014 = 15.603$.

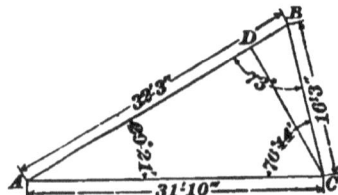

FIG. 11.

$$BC = \frac{CD}{\sin B} = \frac{15.603}{\sin 73° 54' 42'} = 16.24 \text{ feet} = 16 \text{ ft. 3 in.}$$

$$BD = \frac{DC}{\tan B} = \frac{15.603}{\tan 73° 54' 42'} = 4.5 \text{ feet.}$$

$$AB = AD + DB = 27.747 + 4.5 = 32.247 = 32 \text{ ft. 3 in.}$$

Ans. $\begin{cases} AB = 32 \text{ ft. 3 in.} \\ BC = 16 \text{ ft. 3 in.} \\ B = 73° 54' 42'. \end{cases}$

(320) In Fig. 8, problem 301, AB is the side of a regular decagon; then, the angle $COB = \frac{1}{20}$ of 360°, or 18°. To find the side CB, we have $CB = OB \times \sin 18°$, or $CB = 9.75 \times .30902 = 3.013$ inches.   Since $AB = 2 \times CB$, $AB = 2 \times 3.013$, or 6.026 inches, which multiplied by 10, the number of sides, equals 60.26 inches.   Ans.

(321) Perimeter of circle equals $2 \times 9.75 \times 3.1416$, or 61.26 inches. $61.26 - 60.26 = 1$ inch, the difference in their perimeters.   Ans.

In order to find the area of the decagon, we must first find the length of the perpendicular CO (see Fig. 8 in answer to question 301); $CO = OB \times \cos 18°$, or $CO = 9.75 \times .95106 = 9.273$.   Area of triangle $AOB = \frac{1}{2} \times 9.273 \times 6.026$, or 27.939, which multiplied by 10, the number of triangles in the decagon, equals 279.39 square inches.   Area of the circle $= 3.1416 \times 9.75 \times 9.75$, or 298.65 square inches.

$298.65 - 279.39 = 19.26$ square inches difference.   Ans.

(322) The diameter of the circle equals $\sqrt{\frac{89.42}{.7854}} = \sqrt{113.8528}$, or 10.67 inches.   Ans.

The circumference equals 10.67×3.1416, or 33.52 inches. Ans.

In a regular hexagon inscribed in a circle, each side is equal to the radius of the circle; therefore, $\frac{10.67}{2} = 5.335$ inches is the length of a side.   Ans.

(323)   Angle $m\,O\,B = \frac{1}{16}$ of 360°, or $22\frac{1}{2}°$. $m\,O = \frac{1}{2}$ of $m\,n = \frac{1}{2}$ of 2, or 1 inch.   (See Fig. 12).

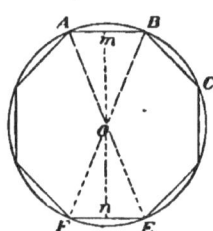

FIG. 12.

Side $m\,B = O\,m \times \tan 22\frac{1}{2}°$, or $m\,B = 1 \times .41421 = .41421$.

$A\,B = 2\,m\,B$; therefore, $A\,B = .82842$ inch.

Area of $A\,O\,B = \frac{1}{2} \times .82842 \times 1 = .41421$ square inch, which, multiplied by 8, the number of equal triangles, equals 3.31368 square inches.

Wt. of bar equals $3.31368 \times 10 \times 12 \times .282$, or 112 pounds 2 ounces.   Ans.

(324) $16 \times 16 \times 16 \times \frac{1}{6} \times 3.1416 = 2{,}144.66$ cu. in. equals the volume of a sphere 16 inches in diameter.

$12 \times 12 \times 12 \times \frac{1}{6} \times 3.1416 = 904.78$ cu. in. equals the volume of a sphere 12 inches in diameter.

The difference of the two volumes equals the volume of the spherical shell, and this multiplied by the weight per cubic inch equals the weight of the shell.   Hence, we have $(2{,}144.66 - 904.78) \times .261 = 323.61$ lb.   Ans.

(325)   The circumference of the circle equals $\frac{5\frac{11}{12} \times 360}{27}$, or 72.0833 inches.   The diameter, therefore, equals $\frac{72.0833}{3.1416}$, or 22.95 inches.

(326)   The number of square inches in a figure 7 inches square equals 7×7 or 49 square inches. $49 - 7 = 42$ square inches difference in the two figures.

$\sqrt{7} = 2.64$ inches is the length of side of square containing 7 square inches.   The length of one side of the other square equals 7 inches.

(327) (*a*) $17\frac{1}{64}$ inches $= 17.016$ inches.

Area of circle $= 17.016^2 \times .7854 = 227.41$ sq. in. **Ans.**

Circumference $= 17.016 \times 3.1416 = 53.457$ inches.

$$16° \; 7' \; 21'' = 16.1225°.$$

(*b*) Length of the arc $= \dfrac{16.1225 \times 53.457}{360} = 2.394$ inches.

Ans.

(328) Area $= 12 \times 8 \times .7854 = 75.4$ sq. in. **Ans.**

Perimeter $= (12 \times 1.82) + (8 \times 1.315) = 32.36$ in. **Ans.**

(329) Area of base $= \dfrac{1}{4} \times 3.1416 \times 7 \times 7 = 38.484$ sq. in.

The slant height of the cone equals $\sqrt{11^2 + 3\frac{1}{2}^2}$, or 11.54 in.

Circumference of base $= 7 \times 3.1416 = 21.9912$.

Convex area of cone $= 21.9912 \times \dfrac{11.54}{2} = 126.927$.

Total area $= 126.927 + 38.484 = 165.41$ square inches.

Ans.

(330) Volume of sphere equals $10 \times 10 \times 10 \times \dfrac{1}{6} \times 3.1416 = 523.6$ cu. in.

Area of base of cone $= \dfrac{1}{4} \times 3.1416 \times 10 \times 10 = 78.54$ sq. in.

$\dfrac{3 \times 523.6}{78.54} = 20$ inches, the altitude of the cone. **Ans.**

(331) Volume of sphere $= \dfrac{1}{6} \times 3.1416 \times 12 \times 12 \times 12 = 904.7808$ cu. in.

Area of base of cylinder $= \dfrac{1}{4} \times 3.1416 \times 12 \times 12 = 113.0976$ sq. in.

Height of cylinder $= \dfrac{904.7808}{113.0976} = 8$ inches. **Ans.**

(332) (*a*) Area of the triangle equals $\dfrac{1}{2} A C \times B D$, or $\dfrac{1}{2} \times 9\frac{1}{2} \times 12 = 57$ square inches. **Ans.**

(*b*) See Fig. 13. Angle $B\,A\,C = 79°\,22'$; angle $A\,B\,D$ $= 90° - 79°\,22' = 10°\,38'$. Side $A\,B = B\,D$ $\div \sin 79°\,22' = 12 \div .98283 = 12.209$ inches.

Side $A\,D = B\,D \times \tan 10°\,38' = 12 \times$ $.18775 = 2.253$ inches.

Side $D\,C = A\,C - A\,D = 9.5 - 2.253 = 7.247$ inches.

Side $B\,C = \sqrt{D\,B^2 + D\,C^2} = \sqrt{12^2 + 7.247^2} = \sqrt{196.519} = 14.018$ inches.

Perimeter of triangle equals $A\,B + B\,C + A\,C = 12.209 + 14.018 + 9.5 = 35.73$ inches. Ans.

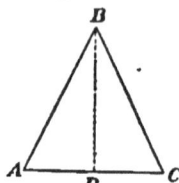

(333) The diagonal divides the trapezium into two triangles; the sum of the areas of these two triangles equals the area of the trapezium, which is, therefore,

$$\frac{11 \times 7}{2} + \frac{11 \times 4\frac{1}{4}}{2} = 61\frac{7}{8} \text{ square inches.}\quad \text{Ans.}$$

(334) Referring to Fig. 17, problem 350, we have $O\,A$ or $O\,B = \dfrac{10}{2}$ or 5 inches, and $A\,B = 6\frac{3}{4}$ inches.

$\operatorname{Sin} C\,O\,B = \dfrac{C\,B}{O\,B} = \dfrac{6\frac{3}{4} \div 2}{5} = .675$; therefore, angle $C\,O\,B = 42°\,27'\,14.3''$.

Angle $A\,O\,B = (42°\,27'\,14.3'') \times 2 = 84°\,54'\,28.6''$. Ans.
$C\,O = O\,B \times \cos C\,O\,B = 5 \times .73782 = 3.6891$.

Area of sector $= 10^2 \times .7854 \times \dfrac{84°\,54'\,28.6''}{360°} = 18.524 \text{ sq.in.}$

Area of triangle $= \dfrac{6.75 \times 3.6891}{2} = 12.450$ sq. in.

$18.524 - 12.450 = 6.074$ sq. in., the area of the segment. Ans.

(335) Convex area $=$

$\dfrac{\text{perimeter of base} \times \text{slant height}}{2} = \dfrac{63 \times 17}{2} = 535.5$ square inches. Ans.

(336) See Fig. 14. Area of lower base
$= 18^2 \times .7854 = 254.469$ sq. in.

Area of upper base $= 12^2 \times .7854 = 113.0976$ sq. in.

$GE = BG - AF = 9 - 6$ or 3 inches.

Slant height $FG = \sqrt{GE^2 + EF^2} = \sqrt{3^2 + 14^2} = 14.32$ inches.

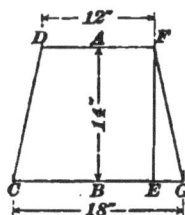
FIG. 14.

Convex area =
$$\frac{\text{circumference of upper base} + \text{circumference of lower base}}{2} \times$$

slant height, or convex area $= \dfrac{37.6992 + 56.5488}{2} \times 14.32 = 674.8156$ sq. in.

Total area $= 674.8156 + 254.469 + 113.0976 = 1,042.3$ sq. in. Ans.

Volume = (area of upper base + area of lower base + $\sqrt{\text{area of upper base} \times \text{area of lower base}}) \times \frac{1}{3}$ of the altitude =

$(113.0976 + 254.4696 + \sqrt{113.0976 \times 254.4696})\dfrac{14}{3} = 2,506.84$ cubic inches. Ans.

(337) Area of surface of sphere 27 inches in diameter $= 27^2 \times 3.1416 = 2,290.2$ sq. in. Ans.

(338) Volume of each ball $= \dfrac{10}{.261} = 38.3142$ cu. in.

Diameter of ball $= \sqrt[3]{\dfrac{38.3142}{.5236}} = 4.18$ inches. Ans.

(339) Area of end $= 19^2 \times .7854 = 283.5294$ sq. in. Volume $= 283.5294 \times 24 = 6,804.7056$ cubic inches $= 3.938$ cubic feet. Ans.

(340) Given $BI = 2$ inches and $HI = IK = \dfrac{14}{2} = 7$ inches to find the radius.

$$BI : HI :: HI : AI, \text{ or } 2 : 7 :: 7 : AI;$$

therefore, $\qquad AI = \dfrac{49}{2} = 24\frac{1}{2}$ inches.

$$AB = AI + BI = 24\frac{1}{2} + 2 = 26\frac{1}{2} \text{ inches.}$$

$$\text{Radius} = \frac{AB}{2} = \frac{26\frac{1}{2}}{2} = 13\frac{1}{4} \text{ inches. Ans.}$$

(341) (*a*) Area of piston = $19^2 \times .7854 = 283.529$ sq. in., or 1.9689 square feet.

Length of stroke plus the clearance = 1.14×2 ft. (24 in. = 2 ft.) = 2.28 ft.

$1.9689 \times 2.28 = 4.489$ cubic feet, or the volume of steam in the small cylinder.   Ans.

(*b*) Area of piston = $31^2 \times .7854 = 754.7694$ sq. in., or 5.2414 square feet.

Length of stroke plus the clearance = $1.08 \times 2 = 2.16$ ft.

$5.2414 \times 2.16 = 11.321$ cubic feet, or the volume of steam in the large cylinder.   Ans.

(*c*) Ratio $= \dfrac{11.321}{4.489}$, or 2.522:1.   Ans.

(342). (*a*) Area of cross-section of pipe = $8^2 \times .7854 = 50.2656$ sq. in.

$$\text{Volume of pipe} = \frac{50.2656 \times 7}{144} = 2.443 \text{ cu. ft.}\quad \text{Ans.}$$

(*b*) Ratio of volume of pipe to volume of small cylinder $= \dfrac{2.443}{4.489}$, or 0.544:1.   Ans.

(343) (*a*) In Fig. 15, given $OB = \dfrac{16}{2}$ or 8 inches, and $OA = \dfrac{13}{2}$ or $6\dfrac{1}{2}$ inches, to find the volume, area and weight: ·

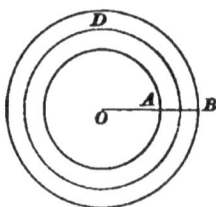

FIG. 15.

Radius of center circle equals $\dfrac{8 + 6.5}{2}$ or $7\dfrac{1}{4}$ inches.   Length of center line = $2 \times 3.1416 \times 7\dfrac{1}{4} = 45.5532$ inches.

The radius of the inner circle is $6\dfrac{1}{2}$ inches, and of the outer circle 8 inches; therefore, the diameter of the cross-section on the line AB is $1\dfrac{1}{2}$ inches.

Then, the area of the ring is $1\dfrac{1}{2} \times 3.1416 \times 45.553 = 214.665$ square inches.   Ans.

Diameter of cross-section of ring $= 1\frac{1}{2}$ inches.

Area of cross-section of ring $=\left(1\frac{1}{2}\right)^{2}\times.7854 = 1.76715$ sq. in.   Ans.

Volume of ring $= 1.76715 \times 45.553 = 80.499$ cu. in.   Ans.

(*b*)   Weight of ring $= 80.499 \times .261 = 21$ lb.   Ans.

(**344**)   The problem may be solved like the one in Art. **790.** A quicker method of solution is by means of the principle given in Art. **826.**

(**345**)   The convex area $= 4 \times 5\frac{1}{4} \times 18 = 378$ sq. in.   Ans.

Area of the bases $= 5\frac{1}{4} \times 5\frac{1}{4} \times 2 = 55.125$ sq. in.

Total area $= 378 + 55.125 = 433.125$ sq. in.   Ans.

Volume $= \left(5\frac{1}{4}\right)^{2} \times 18 = 496.125$ cu. in.   Ans.

(**346**)   In Fig. 16, $O\,C = \dfrac{A\,C}{\tan 30°}.$ $\Bigl(\dfrac{1}{6}$ of $360° = 60°$, and since $A\,O\,C = \dfrac{1}{2}$ of $A\,O\,B,\ A\,O\,C = 30°\Bigr).$

$O\,C = \dfrac{6}{.57735} = 10.392.$

Area of $A\,O\,B = \dfrac{12 \times 10.392}{2} = 62.352$ square feet.

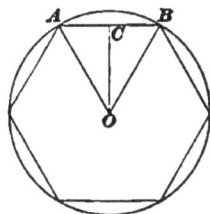

FIG. 16.

Since there are 6 equal triangles in a hexagon, then the area of the base $= 6 \times 62.352$, or 374.112 square feet.

Perimeter $= 6 \times 12$, or 72 feet.

Convex area $= \dfrac{72 \times 37}{2} = 1,332$ sq. ft.   Ans.

Total area $= 1,332 + 374.112 = 1,706.112$ sq. ft.   Ans.

(**347**)   Area of the base $= 374.112$ square feet, and altitude $= 37$ feet. Since the volume equals the area of the base multiplied by $\dfrac{1}{3}$ of the altitude, we have

Volume $= 374.112 \times \dfrac{37}{3} = 4,614$ cubic feet.

(348)   Area of room $= 15 \times 18$ or 270 square feet.

One yard of carpet 27 inches wide will cover $3 \times 2\frac{1}{4}$ (27

inches $= 2\frac{1}{4}$ ft.) $= 6\frac{3}{4}$ sq. ft.   To cover 270 sq. ft., it will

take $\frac{270}{6\frac{3}{4}}$, or 40 yards.   Ans.

(349)   Area of ceiling $= 16 \times 20 = 320$ square feet.
Area of end walls $= 2(16 \times 11) = 352$ square feet.
Area of side walls $= 2(20 \times 11) = \underline{440}$ square feet.
Total area $= \overline{1,112}$ square feet.

From the above number of square feet, the following deductions are to be made:

Windows $= 4(7 \times 4) = 112$ square feet.
Doors $= 3(9 \times 4) = 108$ square feet.

Baseboard less the width of the three doors

equals $(72' - 12') \times \frac{6}{12} = 30$ square feet.

Total No. of feet to be deducted $= 250$ square feet.

Number of square feet to be plastered, then, equals 1,112

$- 250$, or 862 square feet, or $95\frac{7}{9}$ square yards.   Ans.

(350)   Given $A\,B = 6\frac{7}{8}$ inches, and $O\,B = O\,A = \frac{10}{2}$ or

5 inches, Fig. 17, to find the area of the sector.

Area of circle $= 10' \times .7854 = 78.54$ square inches.

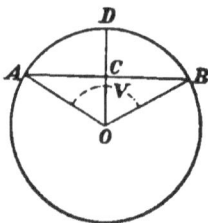

FIG 17.

$\text{Sin}\ A\ O\ C = \dfrac{A\,C}{O\,A} = \dfrac{6\frac{7}{8} \div 2}{5} = .6875$;

therefore, $A\ O\ C = 43°\ 26'$.

$A\ O\ B = 2 \times A\ O\ C = 2 \times 43°\ 26' = 86°\ 52' = 86.8666°$.

$\dfrac{86.8666}{360} \times 78.54 = 18.95$ square inches.

Ans.

(351) Area of parallelogram equals

$$7 \times 10\frac{3}{4} \ (120 \text{ inches} = 10\frac{3}{4}\text{ ft.}) = 75\frac{1}{4} \text{ sq. ft.} \quad \text{Ans.}$$

(352) (a) See Art. 778.

Area of the trapezoid $= \dfrac{15\frac{7}{12} + 21\frac{11}{12}}{2} \times 7\frac{2}{3} = 143.75$ sq. ft.

Ans.

(b) In the equilateral triangle ABC, Fig. 18, the area, 143.75 square feet, is given to find a side. Since the triangle is equilateral all the angles are equal to $\frac{1}{3}$ of 180° or 60°. In the triangle $ABD = ADC$, we have $AD = AB \times \sin 60°$. The area of any triangle is equal to one-half the product of the base by the altitude, therefore, $\dfrac{BC \times AD}{2} = 143.75$. $BC = AB$ and $AD = AB \times \sin 60°$; then, the above becomes

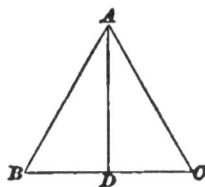

Fig. 18.

$$\frac{AB \times AB \sin 60°}{2} = 143.75,$$

$$\text{or} \quad \frac{AB^2 \times .86603}{2} = 143.75,$$

$$\text{or} \quad AB^2 = \frac{2 \times 143.75}{.86603}.$$

Therefore, $AB = \sqrt{\dfrac{287.50}{.86603}} = 18$ ft. 2.64 in. Ans.

(353) (a) Side of square having an equivalent area $= \sqrt{143.75} = 11.99$ feet. Ans.

(b) Diameter of circle having an equivalent area $= \sqrt{\dfrac{143.75}{.7854}} = \sqrt{183.0277} = 13\frac{1}{2}$ feet. Ans.

(c) Perimeter of square $= 4 \times 11.99 = 47.96$ ft.

Circumference or perimeter of circle $= 13\frac{1}{2} \times 3.1416 = 42.41$ ft.

Difference of perimeter $= \overline{5.55}$ ft. $=$ 5 feet 6.6 inches. Ans.

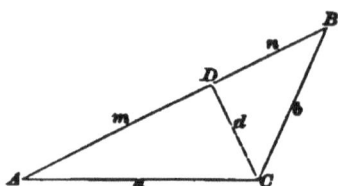

FIG. 19.

(354) In the triangle ABC, Fig. 19,

$AB = 24$ feet,

$BC = 11.25$ feet, and

$AC = 18$ feet.

$$m + n : a + b :: a - b : m - n,$$

or $24 : 29.25 :: 6.75 : m - n$.

$$m - n = \frac{29.25 \times 6.75}{24} = 8.226562.$$

Adding $m + n$ and $m - n$, we have

$$m + n = 24$$
$$m - n = \ 8.226562$$

$$2\,m = 32.226562$$
$$m = 16.113281.$$

Subtracting $m - n$ from $m + n$, we have

$$2n = 15.773438$$
$$n = \ 7.886719.$$

In the triangle ADC, side $AC = 18$ feet, side $AD = 16.113281$; hence, according to Rule **3**, Art. **754**, cos $A = \frac{16.113281}{18} = .89518$, or angle $A = 26° 28' 5''$. In the triangle BDC, side $BD = 7.886719$, and side $BC = 11.25$ ft. Hence, cos $B = \frac{7.886719}{11.25} = .70104$, or angle $B = 45° 29' 23''$.

Angle $C = 180° - (45° 29' 23'' + 26° 28' 5'') = 108° 2' 32''$.

Ans. $\begin{cases} A = 26° 28' \ 5''. \\ B = 45° 29' 23''. \\ C = 108° \ 2' 32''. \end{cases}$

ELEMENTARY MECHANICS.

(355) Use formulas **18** and **8**.

Time it would take the ball to fall to the ground $= t =$

$$\sqrt{\frac{2h}{g}} = \sqrt{\frac{2 \times 5.5}{32.16}} = .58484 \text{ sec.}$$

The space passed through by a body having a velocity of 500 ft. per sec. in .58484 of a second $= S = Vt = 500 \times .58484 = 292.42$ ft.   Ans.

(356) Use formula **7**.

$$\frac{\frac{8.0}{1.2} \times 3.1416 \times 160}{60} = 55.85 \text{ ft. per sec.}   \text{Ans.}$$

(357) $160 \div 60 \times 7 = \frac{8}{21}$ revolution in $\frac{1}{7}$ sec. $360° \times$

$\frac{8}{21} = 137\frac{1}{7}° = 137° 8' 34\frac{2}{7}'$.   Ans.

(358) (*a*) See Fig. 20. $36'' = 3'$. $4 \div 3 = \frac{4}{3} =$ number of revolutions of pulley to one revolution of fly-wheel. $54 \times \frac{4}{3} = 72$ revolutions of pulley and drum per min. $100 \div$ $\left(\frac{18}{12} \times 3.1416 \right) = 21.22$ revolutions of drum to raise elevator 100 ft. $\frac{21.22}{72} \times 60 = 17.68$ sec. to travel 100 ft.   Ans.

(*b*) $21.22 : x :: 30 : 60$, or $x = \frac{21.22 \times 60}{30} = 42.44$ rev.

per min. of drum. The diameter of the pulley divided by

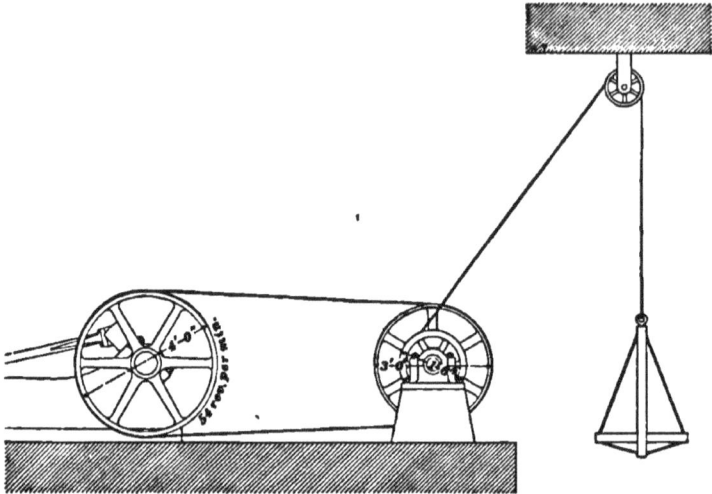

FIG. 20.

the diameter of the fly-wheel $= \frac{3}{4}$, which multiplied by 42.44 $=$ 31.83 revolutions per min. of fly-wheel.   Ans.

(359)   See Arts. **857** and **859.**

(360)   See Art. **861.**

(361)   See Arts. **843** and **871.**

(362)   See Art. **871.**

(363)   See Arts. **842, 886, 887,** etc.

The relative weight of a body is found by comparing it with a given standard by means of the balance.   The absolute weight is found by noting the pull which the body will exert on a spring balance.

The absolute weight increases and decreases according to the laws of weight given in Art. **890;** the relative weight is always the same.

(364)   See Art. **861.**

(365)   See Art. **857.**

(366) See Art. **857.**

(367) If the mountain is at the same height above, and the valley at the same depth below sea-level respectively, it will weigh more at the bottom of the valley.

(368) $\dfrac{31,680}{5,280} = 6$ miles. Using formula **12,** $d^2 : R^2 ::$

$W : w$, we have $w = \dfrac{R^2 W}{d^2} = \dfrac{3,960^2 \times 20,000}{3,966^2} = 19,939.53 +$ lb.

$= 19,939$ lb. $8\frac{1}{2}$ oz. Ans.

(369) Using formula **11,** $R : d :: W : w$, we have

$w = \dfrac{d W}{R} = \dfrac{3,958 \times 20,000}{3,960} = 19,989.89$ lb. $= 19,989$ lb. $14\frac{1}{4}$ oz. Ans.

(370) See Art. **870.**

(371) See Art. **894.**

(372) The velocity which a body may have at the instant the time begins to be reckoned.

(373) Because the man after jumping tends to continue in motion with the same velocity as the train, and the sudden stoppage by the earth causes a shock, the severity of which varies with the velocity of the train.

(374) See Arts. **870** and **871.**

(375) See Art. **872.**

(376) That force which will produce the same final effect upon a body as all the other forces acting together is called the resultant.

(377) (*a*) If a 5-in. line $= 20$ lb., a 1-in. line $= 4$ lb.

$1 \div 4 = \frac{1}{4}$ in. $= 1$ lb. Ans. (*b*) $6\frac{1}{4} \div 4 = 1.5625$ in. $=$

$6\frac{1}{4}$ lb. Ans.

(378) Those forces by which a given force may be

replaced, and which will produce the same effect upon a body.

(379) Southeast, in the direction of the diagonal of a square. See Fig. 21.

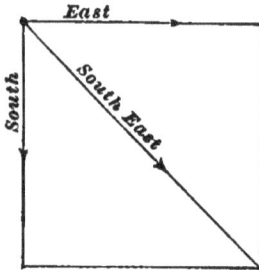

FIG. 21.

(380) $4' 6' = 54'$. $54 \times 2 \times \frac{3}{4} \times$

$.261 = 21.141$ lb. $=$ weight of lever. Center of gravity of lever is in the middle, at a, Fig. 22, 27″ from each end. Consider that the lever has no weight. The center of gravity of the two weights is at b, at a dis-

tance from c equal to $\dfrac{47 \times 54}{47 + 71} = 21.508'' = bc$. Formula **20**, Art. **911.**

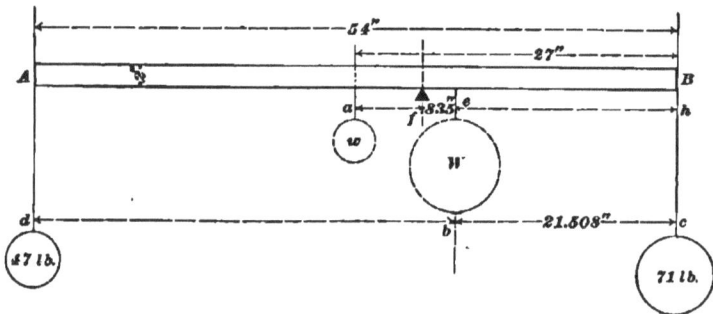

FIG. 22.

Consider both weights as concentrated at b, that is, imagine both weights removed and replaced by the dotted weight W, equal to $71 + 47 = 118$ lb. Consider the weight of the bar as concentrated at a, that is, as if replaced by a weight $w = 21.141$ lb. Then, the distance of the balancing point f, from c, or fc, $= \dfrac{21.141 \times 5.492}{21.141 + 118} = .835'$, since $ac = 27 - 21.508 = 5.492''$. Finally, $fc + ch = fh = .835 + 21.508 = 22.343'' =$ the short arm. Ans. $54 - 22.343 = 31.657'' =$ long arm. Ans.

(381)   See Fig. 23.

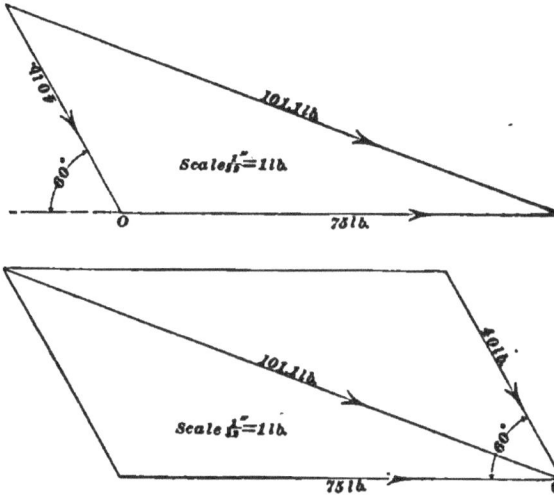

(382)   See Fig. 24.

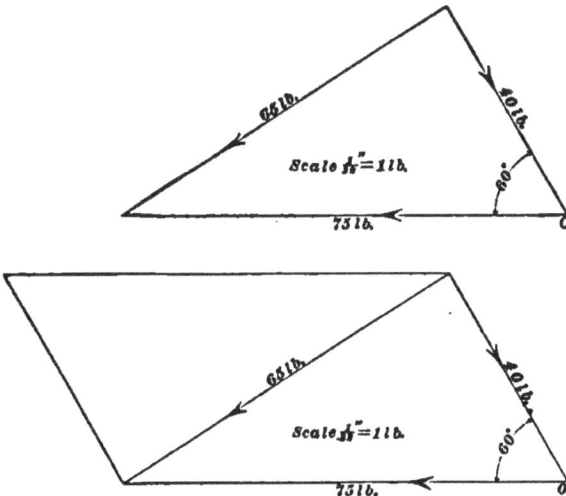

FIG. 24.

(383) $46 - 27 = 19$ lb., acting in the direction of the force of 46 lb.   Ans.

(384)   (*a*) $18 \times 60 \times 60 = 64{,}800$ miles per hour   Ans.

(*b*) $64{,}800 \times 24 = 1{,}555{,}200$ miles.   Ans.

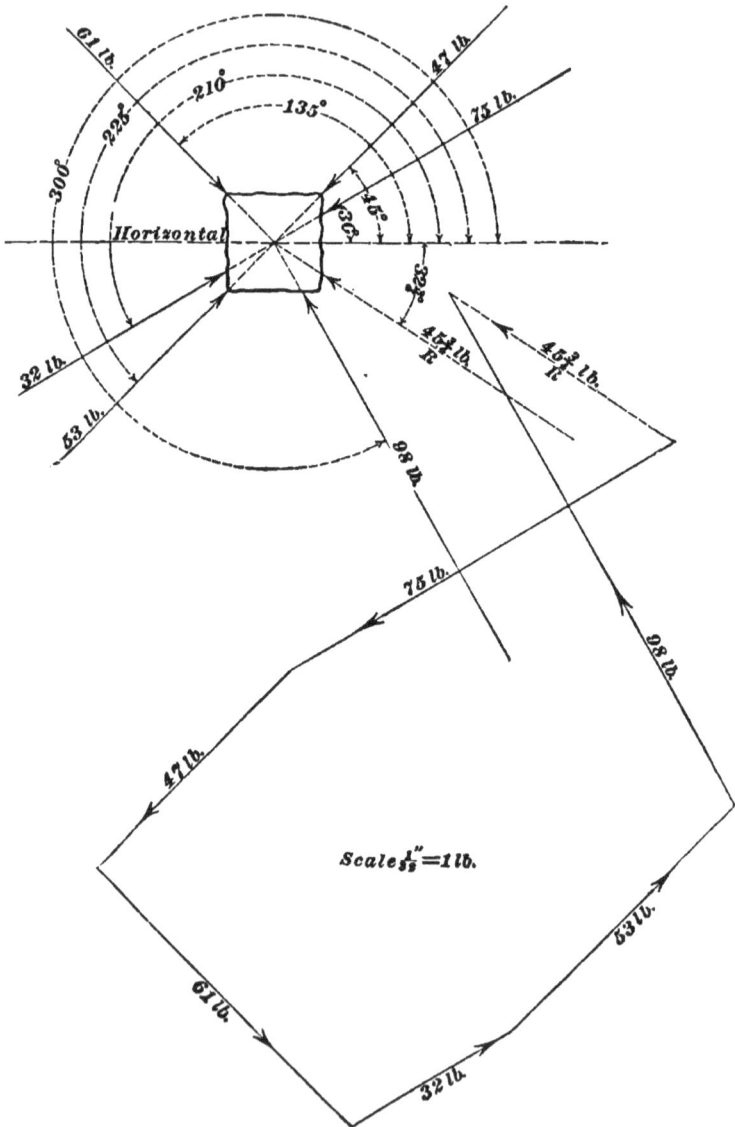

FIG. 25.

(385) (*a*) 15 miles per hour $= \dfrac{15 \times 5,280}{60 \times 60} = 22$ ft. per sec. As the other body is moving 11 ft. per sec., the distance between the two bodies in one second will be $22 + 11 = 33$ ft., and in 8 minutes the distance between them will be $33 \times 60 \times 8 = 15,840$ ft., which, divided by the number of feet in one mile, gives $\dfrac{15,840}{5,280} = 3$ miles. Ans.

(*b*) As the distance between the two bodies increases 33 ft. per sec., then, 825 divided by 33 must be the time required for the bodies to be 825 ft. apart, or $\dfrac{825}{33} = 25$ sec. Ans.

(386) See Fig. 25.

(387) (*a*) Although not so stated, the velocity is evidently considered with reference to a point on the shore. $10 - 4 = 6$ miles an hour. Ans.

(*b*) $10 + 4 = 14$ miles an hour. Ans.

(*c*) $10 - 4 + 3 = 9$, and $10 + 4 + 3 = 17$ miles an hour. Ans.

(388) See Fig. 26.

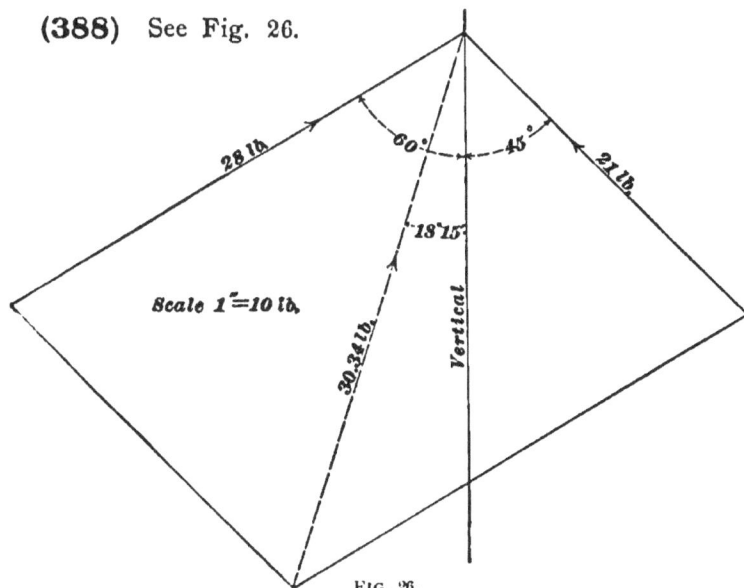

FIG. 26.

(389) See Fig. 27. By rules **2** and **4**, Art. **754,** $b\,c =$ 87 sin 23° = 87 × .39073 = 33.994 lb., $a\,c =$ 87 cos 23° = 87 × .92050 = 80.084 lb.

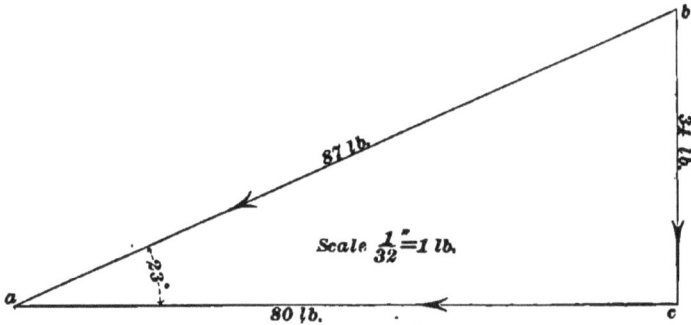

Scale $\frac{1}{32}'' = 1$ lb.

87 lb.

34 lb.

80 lb.

23°

FIG. 27.

(390) See Fig. 28. (*b*) By rules **2** and **4**, Art. **754,** $b\,c = 325$ sin 15° = 325 × .25882 = 84.12 lb. Ans.

(*a*) $a\,c = 325$ cos 15° = 325 × .96593 = 313.93 lb. Ans.

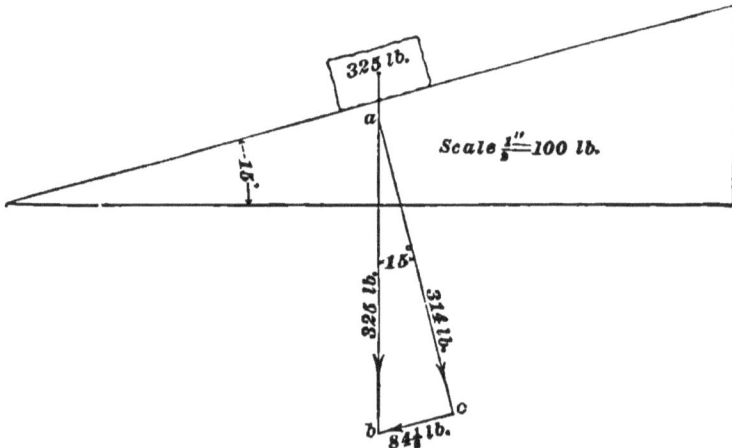

325 lb.

Scale $\frac{1}{8}'' = 100$ lb.

15°

325 lb.

314 lb.

84 lb.

FIG. 28.

(391) Use formula **10.**

$$m = \frac{W}{g} = \frac{125}{32.16} = 3.8868. \quad \text{Ans.}$$

(392) Using formula **10,** $m = \dfrac{W}{g}$, $W = mg = 53.7 \times$ 32.16 = 1,727 lb. Ans.

(393) (*a*) Yes. (*b*) 25. (*c*) 25. Ans.

(394) (*a*) Using formula **12,** $d^2 : R^2 :: W : w$, $d =$ $\sqrt{\dfrac{R^2\,W}{w}} = \sqrt{\dfrac{4,000^2 \times 141}{100}} = 4,749.736$ miles. 4,749.736 − 4,000 = 749.736 miles. Ans.

(*b*) Using formula **11,** $R : d :: W : w$, $d = \dfrac{Rw}{W} =$ $\dfrac{4,000 \times 100}{141} = 2,836.88$ miles.   4,000 − 2,836.88 = 1,163.12 miles. Ans.

(395) (*a*) Use formula **18,**
$$t = \sqrt{\frac{2h}{g}} = \sqrt{\frac{2 \times 5,280}{32.16}} = 18.12 \text{ sec. Ans.}$$

(*b*) Use formula **13,** $v = gt = 32.16 \times 18.12 = 582.74$ ft. per sec., or, by formula **16,** $v = \sqrt{2gh} = \sqrt{2 \times 32.16 \times 5,280} = 582.76$ ft. per sec. Ans.

The slight difference in the two velocities is caused by not calculating the time to a sufficient number of decimal places, the actual value for *t* being 18.12065 sec.

(396) Use formula **25.** Kinetic energy $= Wh = \dfrac{Wv^2}{2g}$.
$$Wh = 160 \times 5,280 = 844,800 \text{ ft.-lb.}$$
$$\frac{Wv^2}{2g} = \frac{160 \times 582.76^2}{2 \times 32.16} = 844,799 \text{ ft.-lb. Ans.}$$

(397) (*a*) Using formulas **15** and **14,**
$h = \dfrac{v^2}{2g} = \dfrac{2,360^2}{2 \times 32.16} = 86,592$ ft. = 16.4 miles. Ans. (*b*)

$t = \dfrac{v}{g} =$ time required to go up or fall back. Hence, total

time $= \dfrac{2v}{g}$ sec. $= \dfrac{2 \times 2,360}{60 \times 32.16} = 2.4461$ min. = 2 min. 26.77 sec. Ans.

(398) 1 hour = 60 min., 1 day = 24 hours; hence, 1 day = 60 × 24 = 1,440 min. Using formula **7**, $V = \dfrac{S}{t}$;

whence, $V = \dfrac{8,000 \times 3.1416}{1,440} = 17.453 +$ miles per min. Ans.

(399) (*a*) Use formula **25**.

Kinetic energy $= \dfrac{Wv^2}{2g} = \dfrac{400 \times 1,875 \times 1,875}{2 \times 32.16} = 21,863,339.55$ ft.-lb. Ans.

(*b*) $\dfrac{21,863,339.55}{2,000} = 10,931.67$ ft.-tons. Ans.

(*c*) See Art. **961**.

Striking force $\times \dfrac{6}{12} = 21,863,339.55$ ft.-lb.,

or striking force $= \dfrac{21,863,339.55}{\frac{6}{12}} = 43,726,679$ lb. Ans.

(400) Using formula **18**, $t = \sqrt{\dfrac{2h}{g}} = \sqrt{\dfrac{2 \times 200}{32.16}} =$ 3.52673 sec., when $g = 32.16$.

$t = \sqrt{\dfrac{2 \times 200}{20}} = 4.47214$ sec., when $g = 20$.

4.47214 − 3.52673 = 0.94541 sec. Ans.

(401). See Art. **910**.

(402) See Art. **963**.

(403) (*a*) See Art. **962**.

$D = \dfrac{m}{V} = \dfrac{W}{gv}$. $v = \dfrac{800}{1,728}$. Hence, $D = \dfrac{W}{gv} = \dfrac{500}{32.16 \times \frac{800}{1728}} = $ 33.582. Ans. (*b*) In Art. **962**, the density of water was found to be 1.941. (*c*) In Art. **963**, it is stated that the specific gravity of a body is the ratio of its density to the density of water. Hence, $\dfrac{33.582}{1.941} = 17.3 =$ specific gravity.

If the weight of water be taken as 62.5 lb. per cu. ft., the specific gravity will be found to be 17.28. Ans.

(404) Assuming that it started from a state of rest, formula **13** gives $v = gt = 32.16 \times 5 = 160.8$ ft. per sec.

(405) Use formulas **17** and **13**. $h = \frac{1}{2} g t^2 =$
$\frac{32.16}{2} \times 3^2 = 144.72$ ft., distance fallen at the end of third second.

$v = g t = 32.16 \times 3 = 96.48$ ft. per sec., velocity at end of third second.

$96.48 \times 6 = 578.88$ ft., distance fallen during the remaining 6 seconds.

$144.72 + 578.88 = 723.6$ ft. = total distance.　Ans.

(406) See Art. **961.**

Striking force $\times \frac{\frac{1}{2}}{12} = 8 \times 8 = 64$.　Therefore, striking force $= \frac{64}{\frac{\frac{1}{2}}{12}} = 1,536$ tons.　Ans.

(407) See Arts. **901** and **902.**

(408) Use formula **19.**

Centrifugal force = tension of string = $.00034 \, W R N^2 =$
$.00034 \times (.5236 \times 4^3 \times .261) \times \frac{15}{12} \times 60^2 = 13.38 +$ lb.　Ans.

(409) $(80^2 - 70^2) \times .7854 \times 26 \times .261 \times \frac{1}{2} = 3,997.2933$ lb., weight of one-half the fly-wheel rim.　Inside radius $= \frac{80 - 10}{2} = 35$ in., or $\frac{35}{12} = 2\frac{11}{12}$ ft.

According to formula **19**, $F = .00034 \, W R N^2 =$
$\frac{.00034 \times 3,997.2933 \times 35 \times 175^2}{12} = 121,394 +$ lb.　Ans.

(410) (*a*) Use formulas **11** and **12**. $R : d :: W : w$, or
$W = \frac{w R}{d} = \frac{1 \times 4,000}{100} = 40$ lb.　Ans.

(*b*)　$d^2 : R^2 :: W : w$, or $w = \frac{4,000^2 \times 40}{4,100^2} = 38.072$ lb. Ans.

(411) See Art. **955.**
$\frac{10,746 \times 354}{10 \times 33,000} = 11.5275$ H. P.　Ans.

(412) Use formula **12.**

$$d^2 : R^2 :: W : w, \text{ or } d = \sqrt{\dfrac{4,000^2 \times 2}{\frac{3}{16}}} = 13,064 \text{ mi., nearly.}$$

$$13,064 - 4,000 = 9,064 \text{ miles.} \quad \text{Ans.}$$

(413) Use formula **18.** $t = \sqrt{\dfrac{2h}{g}} = \sqrt{\dfrac{2 \times 50}{32.16}} =$ 1.7634 sec., nearly.

$$1.7634 \times 140 = 246.876 \text{ ft.} \quad \text{Ans.}$$

(414) $\dfrac{10}{30} = \dfrac{1}{3}$ sec. Use formula **17.**

$$h = \frac{1}{2} g t^2 = \frac{1}{2} \times 32.16 \times \left(\frac{1}{3}\right)^2 = 1.78\frac{2}{3} \text{ ft.} = 1 \text{ ft. } 9.44 \text{ in.}$$
$$\text{Ans.}$$

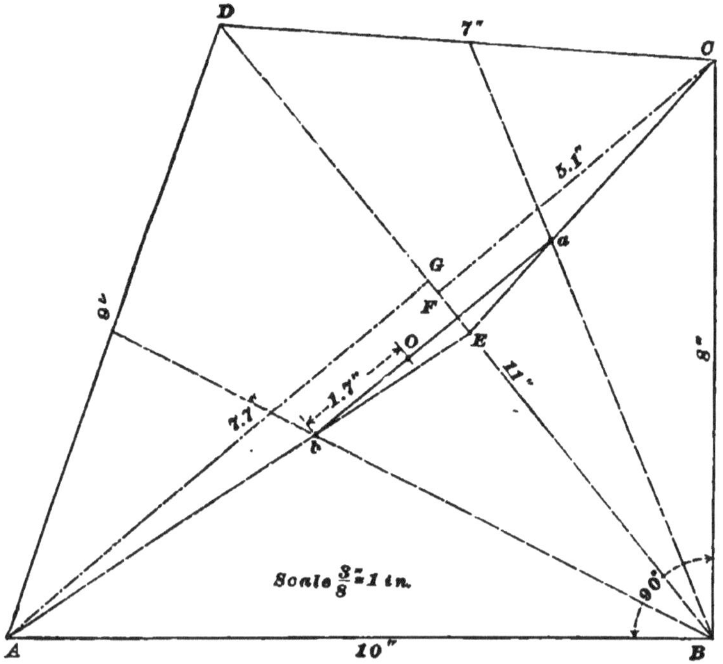

FIG. 29.

(415) See Arts. **906, 907.**

(416) See Arts. **908, 909.**

(**417**) No. It can only be counteracted by another equal couple which tends to revolve the body in an opposite direction.

(**418**) See Art. **914.**

(**419**) Draw the quadrilateral as shown in Fig. 29. Divide it into two triangles by the diagonal $B\ D$. The center of gravity of the triangle $B\ C\ D$ is found to be at a, and the center of gravity of the triangle $A\ B\ D$ is found to be at b (Art. **914**). Join a and b by the line $a\ b$, which, on being measured, is found to have a length of 4.27 inches. From C and A drop the perpendiculars $C\ F$ and $A\ G$ on the diagonal $B\ D$. Then, area of the triangle $A\ B\ D = \frac{1}{2}\ (A\ G \times B\ D)$, and area of the triangle $B\ C\ D = \frac{1}{2}\ (C\ F \times B\ D)$. Measuring these distances, $B\ D = 11'$, $C\ F = 5.1'$, and $A\ G = 7.7'$.

$$\text{Area } A\ B\ D = \frac{1}{2} \times 7.7 \times 11 = 42.35 \text{ sq. in.}$$

$$\text{Area } B\ C\ D = \frac{1}{2} \times 5.1 \times 11 = 28.05 \text{ sq. in.}$$

According to formula **20,** the distance of O, the center of gravity, from b is $\frac{28.05 \times 4.27}{28.05 + 42.35} = 1.7$. Therefore, the center of gravity is on the line $a\ b$ at a distance of $1.7'$ from b.

(**420**) See Fig. 30. The center of gravity lies at the geometrical center of the pentagon, which may be found as follows: From any vertex draw a line to the middle point of the opposite side. Repeat the operation for any other vertex, and the intersection of the two lines will be the desired center of gravity.

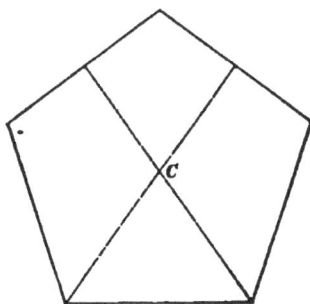

FIG. 30.

(**421**) See Fig. 31. Since any number of quadrilaterals can be drawn with the sides given, any number of answers can be obtained.

Draw a quadrilateral, the lengths of whose sides are equal to the distances between the weights, and locate a weight on each corner. Apply formula **20** to find the distance $C_1 W_1$;

thus, $C_1 W_1 = \dfrac{9 \times 18}{9 + 21} = 5.4''$. Measure the distance $C_1 W_3$;

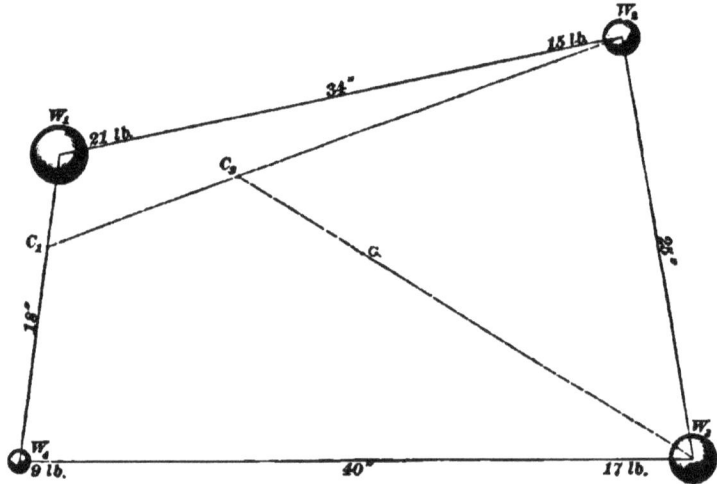

FIG. 31.

suppose it equals say $36''$. Apply the formula again.

$C_1 C_2 = \dfrac{15 \times 36}{15 + (9 + 21)} = 12''$. Measure $C_2 W_3$; it equals say $31.7''$.

Apply the formula again. $C_2 C = \dfrac{17 \times 31.7}{17 + 15 + 9 + 21} = 8.7''$.

C is center of gravity of the combination.

(**422**) Let $A\ B\ C\ D\ E$, Fig. 32, be the outline, the right-angled triangle cut-off being $E\ S\ D$. Divide the figure into two parts by the line $m\ n$, which is so drawn that it cuts off an isosceles right-angled triangle $m\ B\ n$, equal in area to $E\ S\ D$, from the opposite corner of the square.

The center of gravity of $A\,m\,n\,C\,D\,E$ is then at C_1, its geometrical center. $B\,m = 4$ in.; angle $B\,m\,r = 45°$; there-

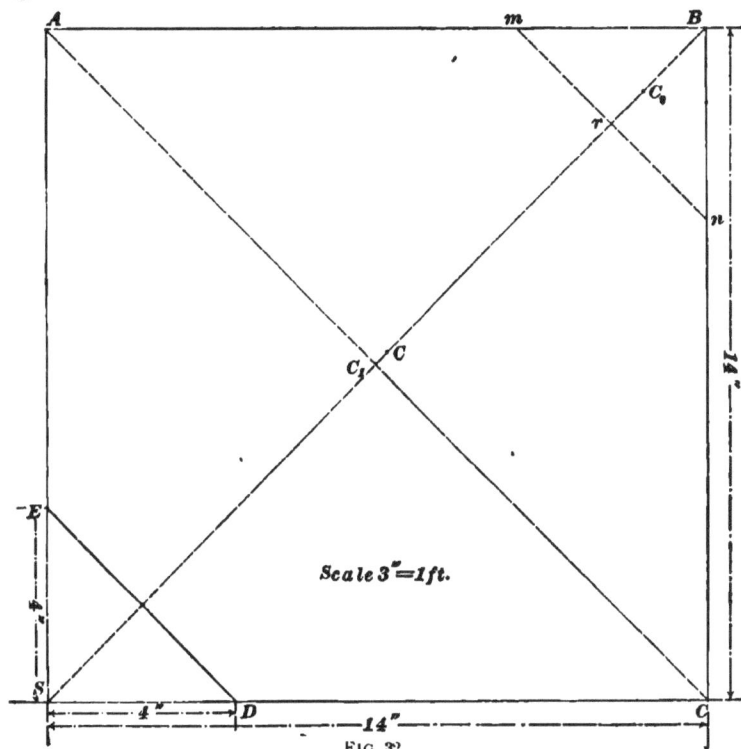

FIG. 32.

fore, $B\,r = B\,m \times \sin B\,m\,r = 4 \times .707 = 2.828$ in. C_2, the center of gravity of $B\,m\,n$, lies on $B\,r$, and $B\,C_2 = \frac{2}{3}B\,r$ $= \frac{2}{3} \times 2.828 = 1.885$ in. $B\,C_1 = A\,B \times \sin\ B\,A\,C_1 = 14 \times$ $\sin 45° = 14 \times .707 = 9.898$ in. $C_1\,C_2 = B\,C_1 - B\,C_2 = 9.898$ $- 1.885 = 8.013$ in.

Area $A\,B\,C\,D\,E = 14^2 - \dfrac{4 \times 4}{2} = 188$ sq. in. Area $m\,B\,n$ $= \dfrac{4 \times 4}{2} = 8$ sq. in. Area $A\,m\,n\,C\,D\,E = 188 - 8 = 180$ sq. in.

The center of gravity of the combined area lies at C, at

a distance from C_1, according to formula **20** (Art. **911**),
equal to $\dfrac{8 \times C_1 C_2}{180 + 8} = \dfrac{8 \times 8.013}{188} = .341$ in. $C_1 C = .341$ in.
$B C = B C_1 - C_1 C = 9.898 - .341 = 9.557$ inches.   Ans.

(423) (*b*) In one revolution the power will have moved through a distance of $2 \times 15 \times 3.1416 = 94.248''$, and the weight will have been lifted $\dfrac{1''}{4}$. The velocity ratio is then

$94.248 \div \dfrac{1}{4} = 376.992.$

$$376.992 \times 25 = 9,424.8 \text{ lb.}   \text{Ans.}$$

(*a*) $9,424.8 - 5,000 = 4,424.8$ lb.   Ans.

(*c*) $4,424.8 \div 9,424.8 = 46.95\%.$ Ans.

(424)   See Arts. **920** and **922.**

(425)   Construct the prism $A\,B\,E\,D$, Fig. 33.   From E, draw the line $E\,F$.   Find the center of gravity of the

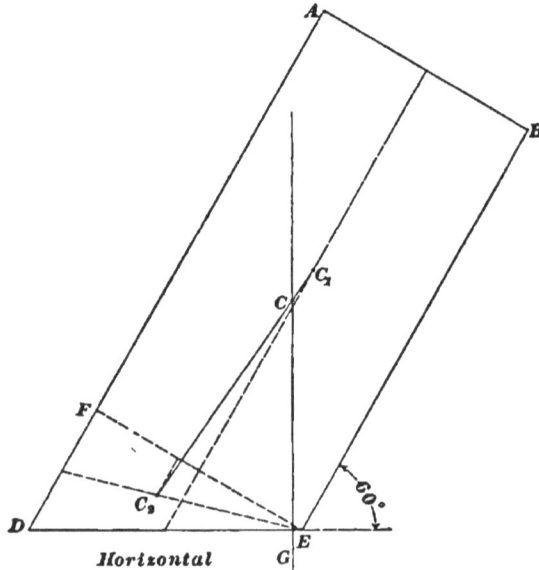

FIG. 33.

rectangle, which is at C_1, and that of the triangle, which is at C_2. . Connect these centers of gravity by the straight line C_1 C_2 and find the common center of gravity of the body by the rule to be at C. Having found this center, draw the line of direction C G. If this line falls within the base, the body will stand, and if it falls without, it will fall.

(**426**) (*a*) 5 ft. 6 in. = 66″. 66 ÷ 6 = 11 = velocity ratio. Ans.

(*b*) 5 × 11 = 55 lb. Ans.

(**427**) 55 × .65 = 35.75 lb. Ans.

(**428**) Apply formula **20.** 5 ft. = 60″. $\dfrac{35 \times 60}{180 + 35} =$ 9.7674 in., nearly, = distance from the large weight. Ans.

(**429**) (*a*) 1,000 ÷ 50 = 20, velocity ratio. Ans. See Art. **945.** (*b*) 10 fixed and 10 movable. ' Ans. (*c*) 50 ÷ 95 = 52.63%. Ans.

(**430**) $P \times$ circumference $= W \times \dfrac{1}{8}$, or 60 × 40 × 3.1416 $= W \times \dfrac{1}{8}$, or $W = 60 \times 40 \times 3.1416 \times 8 = 60{,}318.72$ lb. Since the efficiency of combination is 40%, the tension on the stud would be .40 × 60,318.72 = 24,127.488 lb. Ans.

(**431**) (*a*) $\sqrt{20^2 + 5^2} = 20.616$ ft. = length of inclined plane.

$P \times$ length of plane $= W \times$ height, or $P \times 20.616 = 1{,}580 \times 5$.

$P = \dfrac{1{,}580 \times 5}{20.616} = 383.2$ lb. Ans. (*b*) In the second case, $P \times$ length of base $= W \times$ height, or $P \times 20 = 1{,}580 \times 5$; hence, $P = \dfrac{1{,}580 \times 5}{20} = 395$ lb. Ans.

(**432**) $W \times 2 = 42 \times 6$, or $W = \dfrac{42 \times 6}{2} = 126$ lb.

$126 + 42 = 168$ lb. $168 \times 1 = W' \times 12$, or $W' = \dfrac{168}{12} = 14$ lb.

Ans.

(433) See Fig. 34. $P \times 14 \times 21 \times 19 = 2\frac{1}{2} \times 3\frac{1}{4} \times 2\frac{7}{8}$ $\times$ 725, or

$$P = \frac{2\frac{1}{2} \times 3\frac{1}{4} \times 2\frac{7}{8} \times 725}{14 \times 21 \times 19} = 3.032 \text{ lb.}\quad \text{Ans.}$$

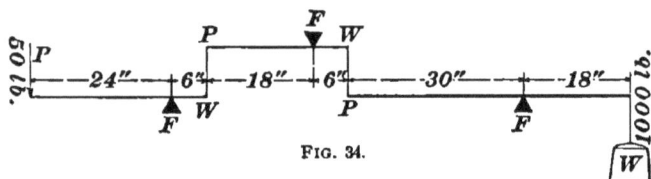

FIG. 34.

(434) See Fig. 35. (a) $35 \times 15 \times 12 \times 20 = 5 \times 3\frac{1}{2} \times$ $3 \times W$, or

$$W = \frac{35 \times 15 \times 12 \times 20}{5 \times 3\frac{1}{2} \times 3} = 2,400 \text{ lb.}\quad \text{Ans.}$$

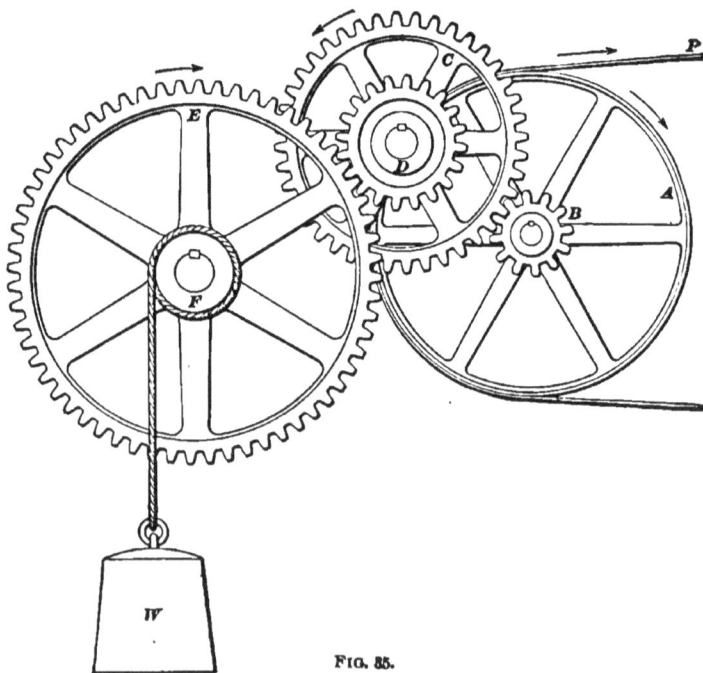

FIG. 35.

(b) $2,400 \div 35 = 68\frac{4}{7} =$ velocity ratio.   Ans.

(c) $1,932 \div 2,400 = .805 = 80.5\%$.   Ans.

(**435**) In Fig. 36, let the 12-lb. weight be placed at A, the 18-lb. weight at B, and the 15-lb. weight at D.

Use formula **20.**

$\dfrac{12 \times 15}{18 + 12} = 6'' =$ distance $C_1 B =$ distance of center of

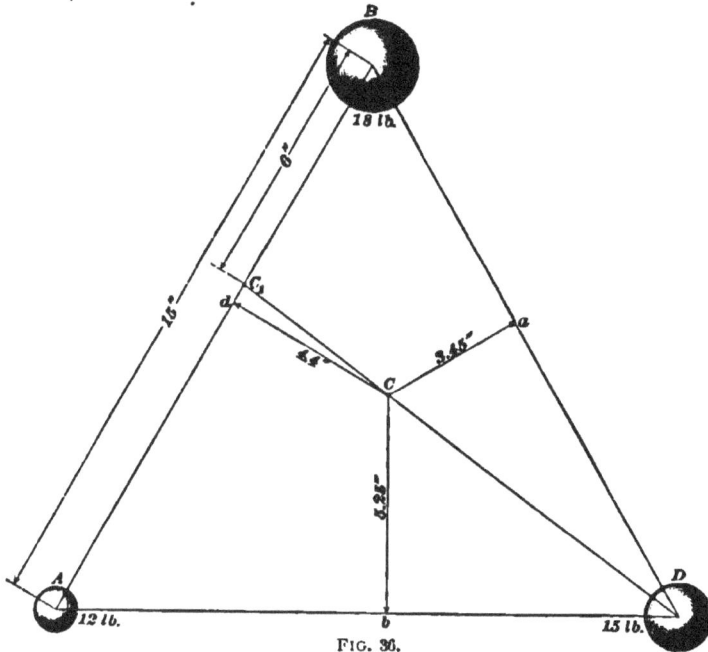

FIG. 36.

gravity of the 12 and 18-lb. weights from B.   Drawing $C_1 D$,

$C_1 C = \dfrac{15 \times C_1 D}{(12 + 18) + 15} = \dfrac{1}{3} C_1 D$.   Measuring the distances of C from BD, DA, and AB, it is found that $Ca = 3.45''$, $Cb = 5.25''$, and $Cd = 4.4''$.   Ans.

(**436**) (a) Potential energy equals the work which the body would do in falling to the ground $= 500 \times 75 = 37,500$ ft.-lb.   Ans.

(*b*)   Using formula **18**, $t = \sqrt{\dfrac{2h}{g}} = \sqrt{\dfrac{2 \times 75}{32.16}} = 2.1597$ sec. $= .035995$ min., the time of falling.

$$\frac{37,500}{33,000 \times .035995} = 31.57 \text{ H. P.}\quad \text{Ans.}$$

(**437**) $127 \div 62.5 = 2.032 = $ specific gravity.   Ans.

(**438**) $\dfrac{62.5}{1,728} \times 9.823 = .35529$ lb.   Ans.

(**439**)   Use formula **21.** $W = \left(\dfrac{2\,PR}{R-r}\right)$,
or $\dfrac{2 \times 60 \times 6.5}{6.5 - 5.75} \times .48 = 499.2$ lb.   Ans.   See Fig. 37.

(**440**)   See Art. **961.**
$$F \times \left(\frac{3}{8} \div 12\right) = \frac{Wv^2}{2g} = \frac{1.5 \times 25^2}{2 \times 32.16}, \text{ or}$$
$$F = \frac{\dfrac{1.5 \times 25^2}{2 \times 32.16}}{\frac{3}{8} \div 12} = 466.42 \text{ lb.}\quad \text{Ans.}$$

(**441**)   (*a*) $2,000 \div 4 = 500 = $ wt. of cu. ft. $500 \div 62.5 = 8 = $ specific gravity.   Ans.

(*b*) $\dfrac{500}{1,728} = .28935$ lb.   Ans.

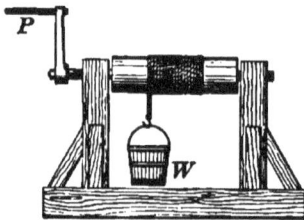

FIG. 37.

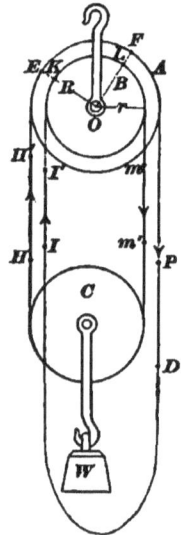

FIG. 38.

(**442**)   See Fig. 38. $14.5 \times 2 = 29.$ $30 \times 29 = W \times 5$, or $W = \dfrac{30 \times 29}{5} = 174$ lb.   Ans.

(**443**) $75 \times .21 = 15.75$ lb.   Ans.

(**444**)   (*a*) $900 \times 150 = 135,000$ ft.-lb.   Ans.
$$\frac{135,000}{15} = 9,000 \text{ ft.-lb. per min.}\quad \text{Ans.}$$

(*b*) $\dfrac{9,000}{33,000} = \dfrac{3}{11}$ H. P.   Ans.

(445) $900 \times .18 \times 2 = 324$ lb. = force required to overcome the friction. $900 + 324 = 1,224$ lb. = total force.

$$\frac{1,224 \times 150}{15 \times 33,000} = .37091 \text{ H. P.} \text{Ans.}$$

(446) $18 \div 88 = .2045.$ Ans.

(447) See Art. **962.** $D = \dfrac{W}{gV} = \dfrac{1,200}{32.16 \times 3} = 12.438.$
Ans.

(448) See Fig. 39. $125 - 47.5 = 77.5$ lb. = downward pressure.

$77.5 \div 4 = 19.375$ lb. = pressure on each support. Ans.

FIG. 39.

(449) See Fig. 40.

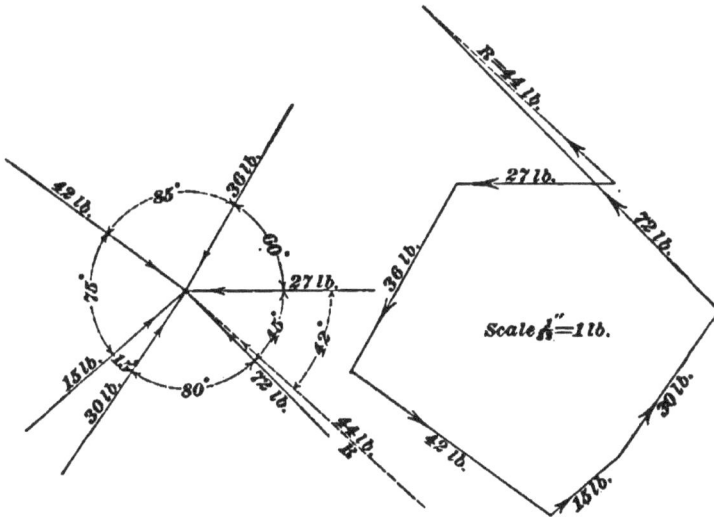

FIG. 40.

(450) See Fig. 41. $4.5 \div 2 = 2.25.$

$$\frac{12}{2.25} \times 6 \times 30 = 960 \text{ lb.} \text{Ans.}$$

(451) (a) $960 \div 30 = 32.$ Ans.

(b) $790 \div 960 = .8229 = 82.29\%.$ Ans.

(452) (*a*) See Fig. 42. $475 + (475 \times .24) = 589$ lb.
Ans.

(*b*) $475 \div 589 = .8064 = 80.64\%$.   Ans.

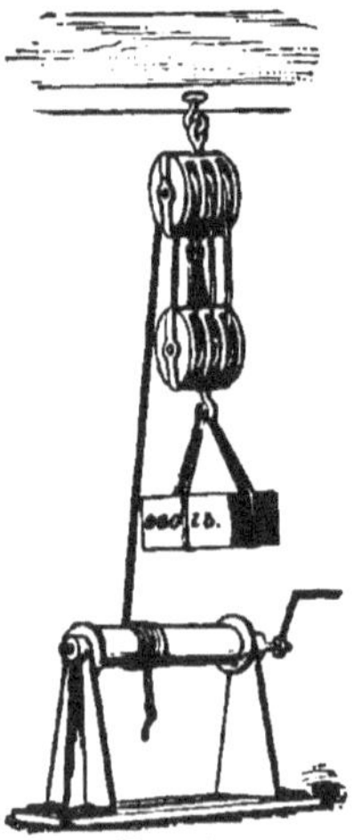

FIG. 41.

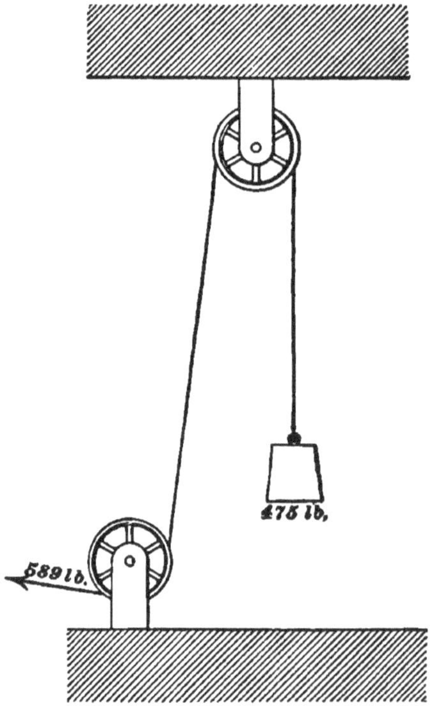

FIG. 42.

(453) (*a*) By formula **23,** $U = FS = 6 \times 25 = 150$
foct-pounds.   Ans.

(*b*) $2\frac{1}{2}$ sec. $= \frac{2\frac{1}{2}}{60} = \frac{1}{24}$ min.

Using formula **24,** Power $= \dfrac{FS}{T} = \dfrac{150}{\frac{1}{24}} = 3,600$ ft.-lb. per
min.   Ans.

HYDROMECHANICS.

---

(454) The area of the surface of the sphere is $20 \times 20 \times 3.1416 = 1,256.64$ sq. in. (See rule, Art. **817.**)

The specific gravity of sea water is 1.026. (See tables of Specific Gravity.) The pressure on the sphere per square inch is the weight of a column of water 1 sq. in. in cross-section and 2 miles long. The total pressure is, therefore, $1,256.64 \times 5,280 \times 2 \times .434 \times 1.026 = 5,908,971$ lb. Ans.

(455) $125 - 83.5 = 41.5$ lb. $=$ loss of weight in water $=$ weight of a volume of water equal to the volume of the sphere. (See Art. **987.**) 1 cu. in. of water weighs .03617 lb.; hence, 41.5 lb. of water must contain $41.5 \div .03617 = 1,147.4$ cu. in. $=$ volume of the sphere. Ans.

NOTE.—It is evident that the specific gravity of the sphere need not be taken into account.

(456) $$Q = \frac{225,000}{60 \times 60} = 62.5 \text{ cu. ft. per min.}$$

Substituting the values given in formula **51,**

$$d = 1.229 \sqrt[5]{\frac{2,800 \times 62.5^2}{26}} = 16.38' +.$$

Substituting this value of d in formula **49,**

$$v_m = \frac{24.51 \times 62.5}{16.38^2} = 5.7095.$$

The value of f (from the table) corresponding to $v_m = 5.7095$ is .0216, using but four decimal places. Hence, applying formula **52,**

$$d = 2.57 \sqrt[r]{\frac{(.0216 \times 2,800 + \frac{1}{4} \times 16.38)62.5^2}{26}} = 16''. \text{ Ans.}$$

(**457**)   (*a*) Area of piston $= \left(\frac{7}{8}\right)^2 \times .7854 = .6013$ sq. in.

Pressure per square inch exerted by piston $= \dfrac{50}{.6013} =$ 83.15 lb.

A column of water 1 foot high and of 1 sq. in. cross-section weighs .434 pound, and therefore exerts a pressure of .434 lb. per sq. in.   The height of a column of water to exert a pressure of 83.15 lb. per sq. in. must be $\dfrac{83.15}{.434} = 191.6$ ft.

Consequently, the water will rise 191.6 ft.   Ans.

The diameter of the hole in the squirt gun has nothing to do with the height of the water, since the pressure per square inch will remain the same, no matter what the diameter may be.

(*b*) Using formula **34**, range $= \sqrt{4\,h\,y}$, we have

$$\sqrt{4\,h\,y} = \sqrt{4 \times 10 \times 191.6} = 87.54 \text{ ft.}   \text{Ans.}$$

(**458**) Use formulas **44** and **43**.

(*b*) $Q_a = .41\,b\,\sqrt{2g}\left[\sqrt{h^3} - \sqrt{h_1^3}\right] =$

$.41 \times \dfrac{30}{12} \times \sqrt{2 \times 32.16}\left[\sqrt{\left(5\frac{1}{2}\right)^3} - \sqrt{\left(3\frac{1}{2}\right)^3}\right] =$ 52.21 cu. ft. per sec.   Ans.

(*a*) Area of weir $= b\,d = 2.5 \times 2 = 5$ sq. ft.

Using formula **43**,

$$v_m = \frac{Q_a}{b\,d} = \frac{52.21}{5} = 10.44 \text{ ft. per sec.}   \text{Ans.}$$

(*c*) To obtain the discharge in gallons per hour (*b*) multiply by 60 × 60 (seconds in an hour) and by 7.48 (gallons in a cu. ft.).   Thus $52.21 \times 60 \times 60 \times 7.48 = 1,405,910.9$ gal. per hour. Ans.

(**459**)   First find the coefficient of friction by using formula **46**, and the table of coefficients of friction.

$$v_m = 2.315\sqrt{\frac{h\,d}{f\,l}} = 2.315\sqrt{\frac{76 \times 7.5}{.025 \times 12,000}} = 3.191 \text{ ft. per sec.}$$

In the table $f = .0243$ for $v_m = 3$ and $.023$ for $v_m = 4$; the difference is $.0013$. $3.191 - 3 = .191$. Then, $1 : .191 :: .0013 : x$, or $x = .0002$. Therefore, $.0243 - .0002 = .0241 = f$. Use formula **50**; substitute in it the value of f here found, and multiply by 60 to get the discharge per minute.

$$Q = .09445\, d^2 \sqrt{\frac{h\,d}{f\,l}} \times 60 =$$

$$.09445 \times 7.5^2 \sqrt{\frac{76 \times 7.5}{.0241 \times 12,000}} \times 60 = 447.6 \text{ gal. per min.}$$

Note.—It will be noticed that the term $\frac{1}{4}d$, in formula **50**, has been omitted. This was done because the length of the pipe exceeded 10,000 times its diameter.

(**460**) (*a*) Use formula **46**.

$$v_m = 2.315 \times \sqrt{\frac{h\,d}{f\,l}} \times 60 =$$

$$2.315 \sqrt{\frac{76 \times 7.5}{.0241 \times 12,000}} \times 60 = 195 \text{ ft. per min.} \quad \text{Ans.}$$

(*b*) 447.6 gal. per min. $\div 60 = 7.46$ gal. per sec. $= 1$ **cu. ft. per sec.**, nearly. Ans.

(**461**) See Art. **1005**.

$$v = \sqrt{2\,g\,h} = \sqrt{2 \times 32.16 \times 10} = 25.36 \text{ ft. per sec.} \quad \text{Ans.}$$

(**462**) Use formulas **49** and **47**.

$$v_m = \frac{24.51\,Q}{d^2} = \frac{24.51 \times 42,000}{6.5^2 \times 60 \times 60} = 6.768 \text{ ft. per sec.}$$

$$h = \frac{f\,l\,v_m^2}{5.36\,d} + .0233\,v_m^2 =$$

$$\frac{.021 \times 1,500 \times 6.768^2}{5.36 \times 6.5} + .0233 \times 6.768^2 = 42.48 \text{ ft.} \quad \text{Ans.}$$

(**463**) (*b*) Area of top or bottom of cylinder equals $20^2 \times .7854 = 314.16$ sq. in. Area of cross-section of pipe $= \left(\frac{3}{8}\right)^2 \times .7854 = .1104$ sq. in. 25 lb. 10 oz. $= 25.625$ lb. $25.625 \div .1104 = 232.11$ lb., pressure per square inch on top or bottom exerted by the weight and piston.

Pressure due to a head of 10 ft. $= .434 \times 10 = 4.34$ lb. per sq. in.

Pressure due to a head of 13 ft. $= .434 \times 13 = 5.64$ lb. per sq. in.

(Since a column of water 1 ft. high exerts a pressure of .434 lb. per sq. in.)

Pressure on the top = pressure due to weight + pressure due to head of 10 ft. $= 232.11 + 4.34 = 236.45$ lb. per sq. in. Ans.

(*a*) Pressure on bottom = pressure due to weight + pressure due to head of 13 ft. $= 232.11 + 5.64 = 237.75$ lb. per sq. in. Ans.

(*c*) Total pressure, or equivalent weight on the bottom = $237.75 \times 314.16 = 74,691.54$ lb. Ans.

(464) $.434 \times 1\frac{1}{2} = .651$ lb., pressure due to the head of water in the cylinder at the center of the orifice.

236.45, pressure per square inch on top, $+ .651 = 237.101$, total pressure per square inch. Area of orifice $= 1^2 \times .7854 = .7854$ sq. in.

$$.7854 \times 237.101 = 186.22 \text{ lb.} \quad \text{Ans.}$$

(465) (*a*) Use formulas **28** and **27**. Sp.Gr. $=$

$$\frac{w}{(W - W') - (W_1 - W_2)} = \frac{11.25}{(91.25 - 41) - (16 \times 5 - 3\frac{1}{8} \times 16)} = .556. \quad \text{Ans.}$$

(*b*) Sp. Gr. $= \dfrac{W}{W - W'} = \dfrac{80}{(16 \times 5) - (3\frac{1}{2} \times 16)} = 2.667.$ Ans.

(466) First find the coefficient of friction. Formula **46** gives

$$v_m = 2.315 \sqrt{\frac{h\,d}{f\,l}} = 2.315 \sqrt{\frac{120 \times 4}{.025 \times 4,000}} = 5.0719 \text{ ft. per sec.}$$

From the table in Art. **1033**, $f = .0230$ for $v_m = 4$, and .0214 for $v_m = 6$. $.0230 - .0214 = .0016$. $5.0719 - 4 = 1.0719$. $2 : 1.0719 :: .0016 : x$, or $x = .0009$. $.0230 - .0009 = .0221 = f$ for $v_m = 5.0719$. Hence,

$$v_m = 2.315 \sqrt{\frac{120 \times 4}{.0221 \times 4,000}} = 5.4 \text{ ft. per sec.} \quad \text{Ans.}$$

(467) Use formulas **46** and **45.**

$$v_m = 2.315 \sqrt{\frac{120 \times 4}{.025 \times 2,000}} = 7.1728 \text{ ft. per sec.}$$

From the table in Art. **1033,** $f = .0214$ for $v_m = 6$, and .0205 for $v_m = 8$. $8 - 6 = 2$.

$.0214 - .0205 = .0009$. $7.1728 - 6 = 1.1728$.

$2 : 1.1728 :: .0009 : x$, or $x = .0005$.

$.0214 - .0005 = .0209 = f$ for $v_m = 7.1728$.

Hence, the velocity of discharge =

$$v_m = 2.315 \sqrt{\frac{120 \times 4}{.0209 \times 2,000 + \frac{1}{2} \times 4}} = 7.8 \text{ ft. per sec. Ans.}$$

(468) (*a*) See Fig.43. Area of cylinder $= 19^2 \times .7854$; pressure, 90 pounds per sq. in. Hence, the total pressure on the piston is $19^2 \times .7854 \times 90 = 25,517.6$ lb. =the load that can be lifted. Ans.

(*b*) The diameter of the pipe has no effect on the load which can be lifted, except that a larger pipe will lift the load faster, since more water will flow in during a given time.

(469) (*a*) $f = .0205$ for $v_m = 8$. Therefore, using formula **47,**

$$h = \frac{.0205 \times 5,280 \times 8^2}{5.36 \times 10} + .0233 \times$$

$8^2 = 130.73$ ft. Ans.

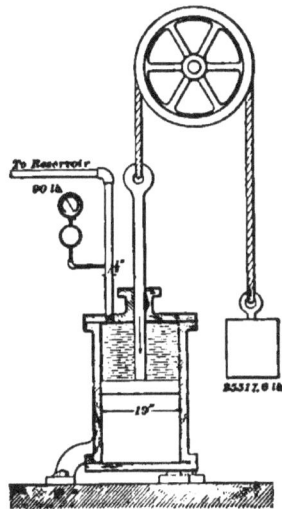

FIG. 43.

(*b*) Using formula **48,** $Q = .0408\,d^2\,v_m = .0408 \times 10^2 \times 8 = 32.64$ gal. per sec. $32.64 \times 60 \times 60 = 117,504$ gal. per hour. Ans.

(470) A column of water 1 in. square and 2.304 ft. high weighs 1 lb.; hence, to produce a pressure of 30 lb. per sq. in. would require a column of water $2.304 \times 30 = 69.12$ ft. high = head. Using formula **36,**

$$v = .98\sqrt{2gh} = .98\sqrt{2 \times 32.16 \times 69.12} = 65.34 \text{ ft. per sec. Ans.}$$

(471) (*a*) 36 in. = 3 ft. A column of water 1 in. square and 1 ft. high weighs .434 lb. .434 × 43 = 18.662 lb. per sq. in., pressure on the bottom of the cylinder. .434 × 40 = 17.36 lb. per sq. in., pressure on the top of the cylinder. Area of base of cylinder = 20' × .7854 = 314.16 sq. in. 314.16 × 18.662 = 5,862.85 lb., total pressure on the bottom. Ans.

(*b*) 314.16 × 17.36 = 5,453.82 lb., total pressure on the top. Ans.

(472) 2 lb. = 32 oz. 32 − 10 = 22 oz. = loss of weight of the bottle in water. 32 + 16 = 48 = weight of bottle and sugar in air. 48 − 16 = 32 oz. = loss of weight of bottle and sugar in water. 32 − 22 = 10 oz. = loss of weight of sugar in water = weight of a volume of water equal to the volume of the sugar. Then, by formula **27,**

$$\text{specific gravity} = \frac{W}{W - W'} = \frac{16}{10} = 1.6. \quad \textbf{Ans.}$$

(473) $33 = \sqrt{2gh}$ (see Art. **1005**), or

$$h = \frac{33^2}{64.32} = 16.931 \text{ ft. per sec. } \textbf{Ans.}$$

(474) (*a*) Use formula **42,** and multiply by 7.48 × 60 × 60 to reduce cu. ft. per sec. to gal. per hour.

$$Q_a = .41 \times \frac{21}{12} \times \sqrt{2 \times 32.16 \times \left(\frac{15}{12}\right)^3} \times 7.48 \times 60 \times 60 =$$

$$216,551 \text{ gal. per hr. } \textbf{Ans.}$$

(*b*) By formula **43,**

$$v_m = \frac{Q_a}{b\,d} = \frac{.41 \times \frac{21}{12} \times \sqrt{2 \times 32.16 \times \left(\frac{15}{12}\right)^3}}{\frac{21}{12} \times \frac{15}{12}} = 3.676 \text{ ft. per sec. } \textbf{Ans.}$$

(475) $f = .0193$ for $v_m = 12$. Therefore, using formula **47,**

$$h = \frac{f\,l\,v_m^2}{5.36\,d} + .0233\,v_m^2 =$$

$$\frac{.0193 \times 6,000 \times 12^2}{5.36 \times 3} + 0233 \times 12^2 = 1,040.37 \text{ ft. } \textbf{Ans}$$

(476) (*a*) 1 cu. in. of water weighs .03617 lb.

.03617 × 40 = 1.4468 lb. = weight of 40 cu. in. of water = loss of weight of lead in water.

16.4 − 1.447 = 14.953 lb. weight of lead in water.   Ans.

(*b*) 16.4 ÷ 40 = .41 = weight of 1 cu. in. of the lead.

16.4 − 2 = 14.4 lb. = weight of lead after cutting off 2 lb.

14.4 ÷ .41 = 35.122 cu. in. = volume of lead after cutting off 2 lb.

(477) (*a*) See Fig. 44.   13.5 × 9 × .7854 = 95.4261 sq. in., area of base.

.03617 × 20 = .7234 lb. per sq. in., pressure on the base due to the water only.

12 + .7234 = 12.7234 lb., total pressure per square inch on base.

12.7234 × 95.4261 = 1,214.144 lb. Ans.

(*b*) 47 × 12 = 564. lb., total upward pressure. Ans.

(478) (*a*) See Fig. 44. 4 sin 53° = 3.195′, nearly.

20 − 3.195 = 16.805″ = distance of center of gravity of plate below the surface.

.03617 × 16.805 + 12 = 12.60784 lb. per sq. in. = per-

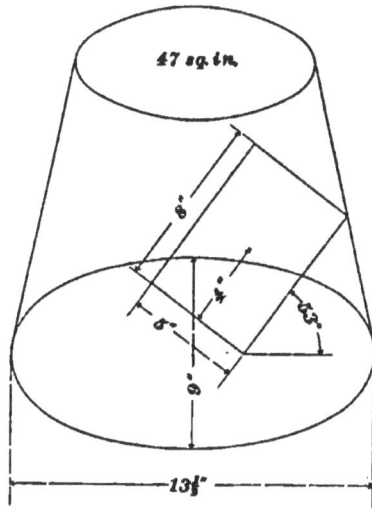

FIG. 44.

pendicular pressure against the plate.   5 × 8 = 40 sq. in., area of plate.

12.60784 × 40 = 504.314 lb. = perpendicular pressure on plate. Ans.

(*b*) 504.314 sin 53° = 402.76 lb. = horizontal pressure on plate. Ans.

(*c*) 504.314 cos 53° = 303.5 lb. = vertical pressure on plate. Ans.

(479) $\dfrac{5^2 \times .7854}{144}$ = area of pipe in square feet. Using formula **31**,

$$Q = A\,v_m = \dfrac{5^2 \times .7854}{144} \times 7.2 = \text{discharge in cu. ft. per sec.}$$

$$\dfrac{5^2 \times .7854}{144} \times 7.2 \times 7.48 = \text{discharge in gal. per sec.}$$

$$\dfrac{5^2 \times .7854}{144} \times 7.2 \times 7.48 \times 60 \times 60 \times 24 = 634,478 \text{ gal. dis-}$$
charged in one day.   Ans.

(480) 38,000 gallons per hour $\dfrac{38,000}{60 \times 60}$ gal. per sec. $= Q.$
Using formula **49**,
$$v_m = \dfrac{24.51\,Q}{d^2} = \dfrac{24.51 \times 38,000}{5.5^2 \times 60 \times 60} = 8.5526 \text{ ft. per sec.}\quad \text{Ans.}$$

(481) Area of $2\frac{1}{2}$-in. circle = 4.9087 sq. in.; area of a 2-in. circle = 3.1416 sq. in. $(4.9087 - 3.1416) \times 12 = 21.2052$ cu. in. of brass.

21.2052 $\times$ .03617 = .767 lb. = weight of an equal volume of water.

6 lb. 5 oz. = 6.3125 lb.   6.3125 $\div$ .767 = 8.23 Sp. Gr. of brass. Ans.

(482) (*b*) A column of water 1 ft. high and 1 in. square weighs .434 lb.   .434 $\times$ 180 = 78.12 lb. per sq. in.   Ans.

(*a*) Projected area of 1 foot of pipe = 6 $\times$ 12 = 72 sq. in. (See Art. **985**.)   72 $\times$ 78.12 = 5,624.64 lb., nearly.   Ans.

(483) Use formula **42**.

(*a*) $Q_a = .41\,b\sqrt{2\,g\,d^3} = .41 \times \dfrac{27}{12} \times \sqrt{2 \times 32.16 \times \left(\dfrac{36}{12}\right)^3} =$
38.44 cu. ft. per sec.   Ans.

(*b*) $Q = \dfrac{Q_a}{.615} = \dfrac{38.44}{.615} = 62.5$ cu. ft. per sec.   Ans.

(484) (*a*) Area of pipe : area of orifice :: 6^2 : 1.5^2; or, area of pipe is 16 times as large as area of orifice. Hence, using formula **35**,

$$v = \sqrt{\frac{2gh}{1 - \frac{a^2}{A^2}}} = \sqrt{\frac{2 \times 32.16 \times 45}{1 - \frac{(1.5^2 \times .7854)^2}{(6^2 \times .7854)^2}}} = 53.9 \text{ ft. per sec. Ans.}$$

(b) $2.304 \times 10 = 23.04$ ft. = height of column of water which will give a pressure of 10 lb. per sq. in. $45 + 23.04 = 68.04$ ft.

$$v = \sqrt{\frac{2 \times 32.16 \times 68.04}{1 - \frac{(1.5^2 \times .7854)^2}{(6^2 \times .7854)^2}}} = 66.28 \text{ ft. per sec. Ans.}$$

(485) Use formula 48.

$Q = .0408 \ d^2 v_m = .0408 \times 6^2 \times 7.5 = 11.016$ gal. per sec. Ans.

(486) $14^2 \times .7854 \times 27 =$ volume of cylinder = volume of water displaced.

$14^2 \times .7854 \times 27 \times .03617 = 150$ lb., nearly, = weight of water displaced.

$(14^2 - 13^2) \times .7854 \times 27 =$ volume of the cylinder walls.

$13^2 \times .7854 \times \frac{1}{4} \times 2 =$ volume of the cylinder ends.

$.261$ lb. = weight of a cubic inch of cast iron, then,

$[(14^2 - 13^2) \times .7854 \times 27 + 13^2 \times .7854 \times \frac{1}{4} \times 2] \times .261 =$

167 lb., nearly, = weight of cylinder. Since weight of cylinder is greater than the weight of the water displaced, it will sink. Ans.

(487) 2 lb. $-$ 1 lb. 5 oz. = 11 oz., weight of water.
1 lb. 15.34 oz. $-$ 1 lb. 5 oz. = 10.34 oz., weight of oil.
$10.34 \div 11 = .94 =$ Sp. Gr. of oil. Ans.

(488) Head $= 41 \div .434 = 94.47$ ft. Using formula 36,
$v = .98\sqrt{2gh} = .98\sqrt{2 \times 32.16 \times 94.47} = 76.39$ ft. per sec. Ans.
This is not the mean velocity, v_m.

(489) (a) Use formula 39.

$$Q_a = .815 \, A \sqrt{2gh}, \text{ or}$$

$$Q_a = .815 \times \frac{1.5^2 \times .7854}{144} \times \sqrt{2 \times 32.16 \times 94.47} \times 60 =$$

46.77 cu. ft. per min. Ans.

(*b*) See Art. **1005.**

The theoretical velocity of discharge is $v = \sqrt{2gh}$, and as $h = 41 \div .434 = 94.47$ ft., we have $v = \sqrt{2 \times 32.16 \times 94.47} = 77.95$ ft. per sec.

Using formula **31,** $Q = A \, v$, and multiplying by 60 to reduce the discharge from cu. ft. per sec. to cu. ft. per min., we have

$$Q = \frac{1.5^2 \times .7854}{144} \times 77.95 \times 60 = 57.39 \text{ cu. ft. per min.} \quad \text{Ans.}$$

(*c*) $\dfrac{Q_a}{Q} = \dfrac{46.77}{57.39} = .815.$ Ans.

(490) (*b*) Use formulas **31** and **38.**

$$Q = A \, v = \frac{1.5^2 \times .7854}{144} \times 77.95 = .9568 \text{ cu. ft. per sec.} \quad \text{Ans.}$$

(*a*) $Q_a = .615 \, Q = .615 \times .9568 = .5884$ cu. ft. per sec. Ans.

(*c*) $\dfrac{Q_a}{Q} = \dfrac{.5884}{.9568} = .615.$ Ans.

(491) (*a*) $9 \times 5 \times .7854 = 35.343$ sq. in. = area of base.

$2^2 \times .7854 = 3.1416$ sq. in. = area of hole.

$\dfrac{3.1416}{35.343} = \dfrac{1}{11.25}$; hence, the area of the base is less than 20 times the area of the orifice, and formula **35** must be used.

$$v = \sqrt{\frac{2gh}{1 - \dfrac{a^2}{A^2}}} = \sqrt{\frac{2 \times 32.16 \times 6}{1 - \dfrac{(2^2 \times .7854)^2}{(9 \times 5 \times .7854)^2}}} = 19.722 \text{ ft. per sec.}$$
                                                                    Ans.

(*b*) $.434 \times 6 = 2.604$, or say 2.6 lb. per sq. in. Ans.

(492) $6 \times 4 \times .7854 = 18.85$ sq. in. = area of upper surface.

$15^2 \times .7854 = 176.715$ sq. in. = area of base.

$\dfrac{132}{18.85} \times 176.715 = 1,237.5$ lb., pressure due to weight on upper surface.

.03617 × 24 × 176.715 = 153.4 lb., pressure due to water in vessel.

1,237.5 + 153.4 = 1,390.9 lb., total pressure.   Ans.

(**493**) Use formula **31** or **32**.

Divide by 60 × 60 to get the discharge in gallons per second, and by 7.48 to get the discharge in cubic feet per second.

$$\text{Area in sq. ft.} = \frac{4^2 \times .7854}{144}.$$

$$v_m = \frac{Q}{A} = \frac{12,000 \times 144}{60 \times 60 \times 7.48 \times 4^2 \times .7854} = 5.106 \text{ ft. per sec.}$$

Ans.

(**494**) A sketch of the arrangement is shown in Fig. 45.

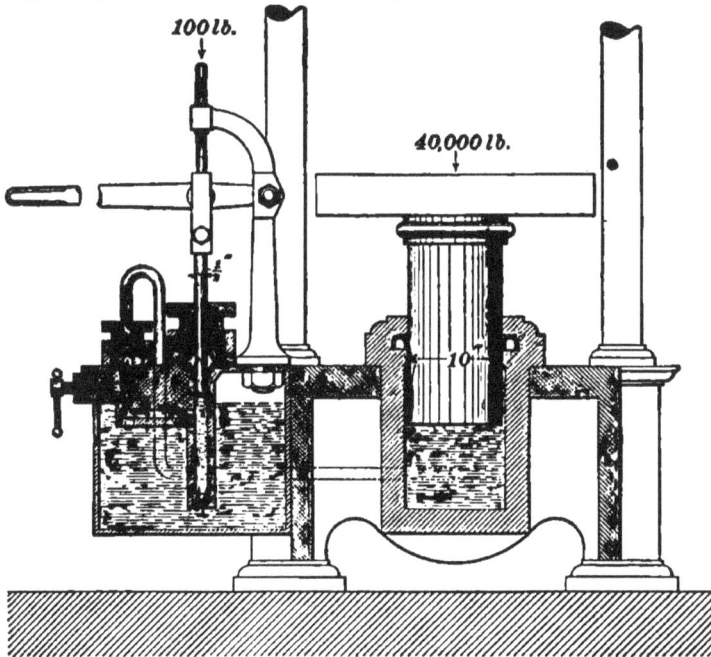

FIG. 45.

(*a*) Area of pump piston $= \left(\frac{1}{2}\right)^2 \times .7854 = .19635$ sq. in.

Area of plunger $= 10^2 \times .7854 = 78.54$ sq. in.

Pressure per square inch exerted by piston $= \dfrac{100}{.19635}$ lb.

Hence, according to Pascal's law, the pressure on the plunger is $\dfrac{100}{.19635} \times 78.54 = 40{,}000$ lb.   Ans.

(b) Velocity ratio $= 1\frac{1}{2} : .00375 = 400 : 1$.   Ans.

(c) According to the principle given in Art. **981,** $P \times 1\frac{1}{2}$ inches $= W \times$ distance moved by plunger, or $100 \times 1\frac{1}{2} = 40{,}000 \times$ required distance; hence, the required distance $= \dfrac{100 \times 1\frac{1}{2}}{40{,}000} = .00375$ in.   Ans.

(495) (a) Use formula **44,** and multiply by 7.48 and 60 to reduce the discharge from cu. ft. per sec. to gal. per min.

$$Q_a = .41\, b \sqrt{2g}\, [\sqrt{h^3} - \sqrt{h_1^3}] \times 60 \times 7.48 =$$

$$.41 \times \frac{14}{12} \times \sqrt{64.32}\left[\sqrt{\left(9 + \frac{20}{12}\right)^3} - \sqrt{9^3}\right] \times 60 \times 7.48 = \quad 13{,}502 \text{ gal.   Ans}$$

(b) In the second case,

$$Q_a = .41 \times \frac{20}{12} \times \sqrt{64.32}\left[\sqrt{\left(9 + \frac{14}{12}\right)^3} - \sqrt{9^3}\right] \times 60 \times 7.48 =$$

$$13{,}323 \text{ gal.   Ans.}$$

(496) (a) Area of weir $= 14 \times 20 \div 144$ sq. ft. Use formula **32,** and divide by 60 × 7.48 to reduce gal. per min. to cu. ft. per sec.

$$v_m = \frac{Q}{A} = \frac{13{,}502 \times 144}{60 \times 7.48 \times 14 \times 20} = 15.47 \text{ ft. per sec.   Ans.}$$

(b) $\quad v_m = \dfrac{13{,}323 \times 144}{60 \times 7.48 \times 14 \times 20} = 15.27$ ft. per sec.   Ans.

(497) (a) See Art. **997.**.

$$W = 2 \text{ lb. } 8\tfrac{1}{2}\text{oz.} = 40.5 \text{ oz.}$$
$$w = \qquad\qquad\ 12 \text{ oz.}$$
$$W' = 1 \text{ lb. } 11 \text{oz.} = 27 \text{ oz.}$$

By formula **30**,

$$\text{Sp. Gr.} = \frac{W - w}{W' - w} = \frac{40.5 - 12}{27 - 12} = \frac{28.5}{15} = 1.9. \quad \text{Ans.}$$

(*b*)   15 oz. $= \frac{15}{16}$ lb. $= .9375$ lb. $.9375 \div .03617 = 25.92$ cu.
in. = volume of water = volume of slate.   Therefore, the
volume of the slate = 25.92 cu. in.   Ans.

(**498**)   In Art. **1019** it is stated that the theoretical
mean velocity is $\frac{2}{3}\sqrt{2gh}$. Hence, $v_m = \frac{2}{3}\sqrt{2 \times 32.16 \times 3} =$
9.26 ft. per sec.   Ans.

(**499**)   (*a*) 4 ft. 9 in. $= 4.75$ ft. $19 - 4.75 = 14.25$.
Range $= \sqrt{4hy} = \sqrt{4 \times 4.75 \times 14.25} = 16.454$ ft.   Ans.

(*b*) $19 - 4.75 = 14.25$ ft.   Ans.

(*c*) $19 \div 2 = 9.5$.   Greatest range $= \sqrt{4 \times 9.5^2} = 19$ ft.
Ans.   (See Art. **1009**.)

(**500**)   Use formulas **46** and **50**.

$$v_m = 2.315\sqrt{\frac{hd}{fl}} = 2.315\sqrt{\frac{25 \times 5}{.025 \times 1,300}} = 4.5397 \text{ ft. per sec.}$$

From the table, $f = .0230$ for $v_m = 4$ and $.0214$ for $v_m = 6$.
$.0230 - .0214 = .0016$. $6 - 4 = 2$.

$4.5397 - 4 = .5397$.   Then, $2 : .5397 :: .0016 : x$, or $x = .0004$.   Hence, $.0230 - .0004 = .0226 = f$ for $v_m = 4.5397$.

$$Q = 60 \times 60 \times .09445 \times 5^2 \times \sqrt{\frac{25 \times 5}{.0226 \times 1,300 + \frac{1}{8} \times 5}} =$$
17,350 gal. per hr.   Ans.

.(**501**)   Obtain the values by approximating to those given
in Art. **1033**. Thus, for $v_m = 2$, $f = .0265$; for $v_m = 3$, $f = .0243$; $.0265 - .0243 = .0022$. $2.37 - 2 = .37$. Hence, $1 : 37 :: .0022 : x$, or $x = .0008$.   Then, $.0265 - .0008 = .0257 = f$
for $v_m = 2.37$.   Ans.

For $v_m = 3$, $f = .0243$; for $v_m = 4$, $f = .0230$; $.0243 - .0230 = .0013$. $3.19 - 3 = .19$.   Hence, $1 : .19 :: .0013 : x$, or $x = .0002$. Then, $.0243 - .0002 = .0241 = f$ for $v_m = 3.19$.   Ans.

For $v_m = 4, f = .0230$; for $v_m = 6, f = .0214$; $.0230 - .0214$ $= .0016$. $5.8 - 4 = 1.8$. $6 - 4 = 2$. Hence, $2 : 1.8 :: .0016 : x$, or $x = .0014$. Then, $.0230 - .0014 = .0216 = f$ for $v_m = 5.8$. Ans.

For $v_m = 6, f = .0214$; for $v_m = 8, f = .0205$; $.0214 - .0205$ $= .0009$. $7.4 - 6 = 1.4$. $8 - 6 = 2$. Hence, $2 : 1.4 = .0009 : x$, or $x = .0006$. Then, $.0214 - .0006 = .0208 = f$ for $v_m = 7.4$. Ans.

For $v_m = 8, f = .0205$; for $v_m = 12, f = .0193$; $.0205 - .0193$ $= .0012$. $9.83 - 8 = 1.83$. $12 - 8 = 4$. Hence, $4 : 1.83 :: .0012 : x$, or $x = .0005$. Then, $.0205 - .0005 = .02 = f$ for $v_m = 9.83$. Ans.

For $v_m = 8, f = .0205$; for $v_m = 12, f = .0193$; $.0205 - .0193 = .0012$. $11.5 - 8 = 3.5$. $12 - 8 = 4$. Hence, $4 : 3.5 :: .0012 : x$, or $x = .0011$. $.0205 - .0011 = .0194 = f$ for $v_m = 11.5$. Ans.

(502) Specific gravity of sea-water is 1.026. Total area of cube $= 10.5^2 \times 6 = 661.5$ sq. in. 1 mile $= 5,280$ ft. Hence, total pressure on the cube $= 661.5 \times 5,280 \times 3.5 \times .434 \times 1.026 = 5,443,383$ lb. Ans.

(503) $19^2 \times .7854 \times 80 = 22,682$ lb. Ans.

PNEUMATICS.

(**504**) The force with which a confined gas presses against the walls of the vessel which contains it.

(**505**) (*a*) $4 \times 12 \times .49 = 23.52$ lb. per sq. in. Ans.

(*b*) $23.52 \div 14.7 = 1.6$ atmospheres. Ans.

(**506**) (*a*) A column of water 1 foot high exerts a pressure of .434 lb. per sq. in. Hence, $.434 \times 19 = 8.246$ lb. per sq. in., the required tension. A column of mercury 1 in. high exerts a pressure of .49 lb. per sq. in. Hence, $8.246 \div .49 = 16.828$ in. = height of the mercury column. Ans.

(*b*) Pressure above the mercury $= 14.7 - 8.246 = 6.454$ lb. per sq. in. Ans.

(**507**) Using formula **53**, $p_1 = \dfrac{p\,v}{v_1} = \dfrac{(14.7 \times 3) \times 1}{2.5} = 17.64$ lb. Ans.

(**508**) (*c*) Using formula **61**,

$$V = \frac{.37052\ W\,T}{p} = \frac{.37052 \times 7.14 \times 535}{22.05} = 64.188 \text{ cu. ft. Ans}$$

($T = 460° + 75° = 535°$, and $p = 14.7 \times 1.5 = 22.05$ lb. per sq. in.)

(*a*) $7.14 \div .08 = 89.25$ cu. ft., the original volume. Ans.

(*b*) If 1 cu. ft. weighs .08 lb., 1 lb. contains $1 \div .08 = 12.5$ cu. ft. Hence, using formula **60**, $p\,V = .37052\,T$, or

$$T = \frac{p\,V}{.37052} = \frac{22.05 \times 12.5}{.37052} = 743.887°.\qquad 743.887 - 460 = 283.887°.\ \text{ Ans.}$$

(**509**) Substituting in formula **59**, $p = 40$, $t = 120$, and $t_1 = 55$,

$$p_1 = 40\left(\frac{460 + 55}{460 + 120}\right) = \frac{40 \times 515}{580} = 35.517 \text{ lb.}\quad \text{Ans.}$$

(**510**) Using formula **61**, $pV = .37052\, WT$, or

$$W = \frac{pV}{.37052\,T}.\quad T = 460° + 60° = 520°.$$

Therefore, $W = \dfrac{14.7 \times 1}{.37052 \times 520} = .076296$ lb.   Ans.

(**511**) $175,000 \div 144 =$ pounds per sq. in.

$$(175,000 \div 144) \div 14.7 = 82.672 = \text{atmospheres.}$$
$$\text{Ans.}$$

(**512**) Extending formula **63** to include 3 gases, we have $PV = p_1v_1 + p_2v_2 + p_3v_3$, or $40 \times P = 1 \times 12 + 2 \times 10 + 3 \times 8$.

Hence, $P = \dfrac{56}{40} = 1.4$ atmos. $= 1.4 \times 14.7 = 20.58$ lb. per sq. in.   Ans.

(**513**) In the last example, $PV = 56$. In the present case, $P = \dfrac{23}{14.7}$ atmos.   Therefore, $V = \dfrac{56}{P} = \dfrac{56}{\frac{23}{14.7}} = 35.79$ cu. ft. Ans.

(**514**) For $t = 280°$, $T = 740°$; for $t = 77°$, $T = 537°$.

$$pV = .37052\,WT, \text{ or } W = \frac{pV}{.37052\,T}.\quad \text{(Formula \textbf{61}.)}$$

$$\text{Weight of hot air} = \frac{14.7 \times 10,000}{.37052 \times 740} = 536.13 \text{ lb.}$$

$$\text{Weight of air displaced} = \frac{14.7 \times 10,000}{.37052 \times 537} = 738.81 \text{ lb.}$$

$738.81 - 536.13 = 202.68.$ $202.68 - 100 = 102.68$ lb. Ans.

(**515**) According to formula **64**,

$$PV = \left(\frac{p_1v_1}{T_1} + \frac{p_2v_2}{T_2}\right)T, \text{ or}$$

$$20 \times 31 = \left(\frac{14.7 \times 13}{533} + \frac{30 \times 18}{513}\right)T = 1.411168\,T.$$

Therefore, $T = \dfrac{20 \times 31}{1.411168} = 439.35°$. Since this is less than 460°, the temperature is $460 - 439.35 = 20.65°$ below zero, or $-20.65°$. Ans.

(**516**) A hollow space from which all air or other gas (or gaseous vapor) has been removed. An example would be the space above the mercury in a barometer.

(**517**) One inch of mercury corresponds to a pressure of .49 lb. per sq. in.

$\dfrac{1}{40}$ inch of mercury corresponds to a pressure of $\dfrac{.49}{40}$ lb. per sq. in. $\dfrac{.49}{40} \times 144 = 1.764$ lb. per sq. ft. Ans.

(**518**) (*a*) $325 \times .14 = 45.5$ lb. $=$ force necessary to overcome the friction. $6 \times 12 = 72' =$ length of cylinder. $72 - 40 = 32 =$ distance which the piston must move. Since the area of the cylinder remains the same, any variation in the volume will be proportional to the variation in the length between the head and piston. By formula **53**, $p\,v = p_1 v_1$. Therefore, $p = \dfrac{p_1 v_1}{v} = \dfrac{14.7 \times 40}{72} = 8.1\frac{2}{3}$ lb. per sq. in. $=$ pressure when piston is at the end of the cylinder. Since there is the atmospheric pressure of 14.7 lb. on one side of the piston and only $8.1\frac{2}{3}$ lb. on the other side, the force required to pull it out of the cylinder is $14.7 - 8.1\frac{2}{3} = 6.5\frac{1}{3}$ lb. per sq. in. Area of piston $= 40' \times .7854 = 1,256.64$ sq. in. Total force $= 1,256.64 \times 6.5\frac{1}{3} = 8,210.05$. Adding the friction, $8,210.05 + 45.5 = 8,255.55$ lb. Ans.

(*b*) Proceeding likewise in the second case, $p\,v = p_1 v_1$, or $p = \dfrac{p_1 v_1}{v} = \dfrac{14.7 \times 40}{6} = 98$ lb. $98 - 14.7 = 83.3$ lb. per sq. in.

$1,256.64 \times 83.3 + 45.5 = 104,723.612$ lb. Ans.

(519) $8.47 =$ original volume $= v_1$. $8.47 - 4.5 = 3.97$ cu. ft. $=$ new volume $= v$. By formula **53**,

$$p_1 = \frac{p\,v}{v_1} = \frac{3.97 \times 38}{8.47} = 17.812 \text{ lb. per sq. in.} \quad \text{Ans.}$$

(520) Original weight $= W = .5$ lb. $= 8$ oz.; new weight $= W_1 = 1$ lb. 6 oz. $= 22$ oz. According to formula **56**,

$$p\,W_1 = p_1\,W, \text{ or } p_1 = \frac{P\,W_1}{W} = \frac{14.7 \times 22}{8} = 40.425 \text{ lb. per}$$
sq. in. Ans.

(521) Applying formula **58**,

$$v_1 = v\left(\frac{460 + t_1}{460 + t}\right) = \frac{4,516}{1,728}\left(\frac{460 + 80}{460 + 260}\right) = 1.96 \text{ cu. ft.} \quad \text{Ans.}$$

(522) According to formula **61**, $p\,V = .37052\,W\,T$, or

$$W = \frac{p\,V}{.37052\,T} = \frac{14.7 \times 1.25 \times 55}{.37052 \times 548} = 4.977 \text{ lb.} \quad \text{Ans.}$$

(523) Using formula **63**, $P\,V = p\,v + p_1\,v_1$, or $P \times 7.5 = 14.7 \times 2 \times 7.5 + 40 \times 7.5$, or $P = 69.4$ lb. per sq. in. Ans.

(524) $48''$, $36''$, and $24'' = 4'$, $3'$, and $2'$, respectively. Hence, $4 \times 3 \times 2 = 24$ cu. ft. $=$ the volume of the block. The block will weigh as much more in a vacuum as the weight of the air it displaces. In example 510, it was found that 1 cu. ft. of air at a temperature of 60° weighed .076296 lb. $.076296 \times 24 + 1,200 = 1,201.83$ lb. Ans.

(525) (*a*) (See Art. **1088.**) $127 + 16 = 143$.

$$\frac{\left(\frac{9}{12}\right)^2 \times .7854 \times 1 \times 125 \times 62.5 \times 143}{33,000} = 14.9563 \text{ H.P.}$$

$$14.9563 \div .75 = 19.942 \text{ H.P.} \quad \text{Ans.}$$

(*b*) Discharge in gallons per hour $=$ volume of cylinder in cu. ft. $\times$ number of strokes per minute $\times 7.48 \times 60 =$

$$\left(\frac{9}{12}\right)^2 \times .7854 \times 125 \times 7.48 \times 60 = 24,784.3 \text{ gal. per hr.} \quad \text{Ans.}$$

(526) In this example, the number of times that the pump delivers water in 1 minute is $100 \div 2 = 50$; in the last example, 125. Hence, the number of gallons discharged per hour in this case will be $24,786 \times \dfrac{50}{125} = 9,914.4$ gal. Ans.

(527) See Art. **1043.**

Pressure in condenser $= \dfrac{30-23}{30} \times 14.7 = 3.43$ lb. per sq. in. Ans.

(528) $144 \times 14.7 = 2,116.8$ lb. per sq. ft. Ans.

(529) $.27 \div 3 = .09 =$ weight of 1 cu. ft. Using formula **56,**

$p\,W_1 = p_1\,W$, or $30\,W_1 = 65 \times .09$. $W_1 = .195$ lb. Ans.

(530) Using formula **61,**

$p\,V = .37052\,W\,T$, or $30 \times 1 = .37052 \times .09 \times T$.

$T = \dfrac{30}{.37052 \times .09} = 899.6°$. $899.6° - 460° = 439.6°$. Ans.

(531) $460° + 32° = 492°$; $460° + 212° = 672°$; $460° + 62° = 522°$, and $460° + (-40°) = 420°$.

(532) Using formula **61,** $p\,V = .37052\,W\,T$, and substituting,

$(14.7 \times 10) \times 4 = .37052 \times 3.5 \times T$, or $T = \dfrac{14.7 \times 10 \times 4}{.37052 \times 3.5} = 453.417°$. $453.417 - 460° = -6.583°$. Ans.

(533) Using formula **63,** $V\,P = v\,p + v_1\,p_1$, we find

$P = \dfrac{15 \times 63 + 19 \times 14.7 \times 3}{25} = 71.316$ lb. Ans.

(534) Using formula **60,** $p\,V = .37052\,T$, or $P = \dfrac{.37052 \times 540}{10} = 20$ lb. per sq. in., nearly. Ans.

(535) One inch of mercury represents a pressure of .49 lb. Therefore, the height of the mercury column is $12.5 \div .49 = 25.51$ in. Ans.

(536) Thirty inches of mercury corresponds to 34 ft. of water. (See Art. **1043.**) Therefore,

$30' : 34$ ft. $:: 27' : x$ ft., or $x = 30.6$ ft. Ans.

A more accurate way is $(27 \times .49) \div .434 = 30.5$ ft.

(537) (a) $30 - 17.5 = 12.5$ in. $=$ original tension of gas in inches of mercury. $30 - 5 = 25$ in. $=$ new tension in inches of mercury.

$VP = vp + v_1 p_1$ (formula **63**), or $6.7 \times 25 = 6.7 \times 12.5 + v_1 \times 30$.

$$v_1 = \frac{6.7 \times 25 - 6.7 \times 12.5}{30} = 2.79\frac{1}{6} \text{ cu. ft. Ans.}$$

(b) To produce a vacuum of 0 inches,

$$v_1 = \frac{6.7 \times 30 - 6.7 \times 12.5}{30} = 3.908 \text{ cu. ft. Ans.}$$

(538) $11 + 25 = 36$, final volume of gas. $2.4 \div 36 = \frac{1}{15}$ lb. Ans.

(539) Using formula **59**,

$$p_1 = p\left(\frac{460 + t_1}{460 + t}\right) = 12 \times \left(\frac{460 + 300}{460 + 60}\right) = 17.54 \text{ lb. per sq. in.}$$
Ans.

(540) $T = 460 + 212 = 672°$. Using formula **61**, $pV = .37052\ WT$, we have $14.7 \times 1 = .37052 \times W \times 672$, or $W = \frac{14.7}{.37052 \times 672} = .059039$ lb. Ans.

(541) (a) $\dfrac{20' \times .7854 \times 32}{1,728} = 5.8178$ cu. ft. $=$ volume of cylinder.

$32 - 26 = 6$ in., length of stroke unfinished.

$$5.8178 \times \frac{6}{32} = 1.0908 \text{ cu. ft. Ans.}$$

(b) By formula **61**, taking the values of p, V, and T at the beginning of stroke,

$pV = .37052\ WT$, or $W = \dfrac{pV}{.37052\ T} = \dfrac{14.7 \times 5.8178}{.37052 \times 535} = .43143$ lb. Ans.

(c) Now, substituting in formula **61** the values of V, W, and T at time of discharge,

$$p = \frac{.37052\ WT}{V} = \frac{.37052 \times .43143 \times 585}{1.0908} = 85.727 \text{ lb. per}$$
sq. in. Ans.

(542) Using formula 63, $VP = vp + v_1 p_1$, or 30×35
$= 19 \times 12 + 21 p_1$, or $p_1 = \dfrac{30 \times 35 - 19 \times 12}{21} = 39.14$ lb. per
sq. in. Ans.

(543) Use formula 64. $PV = \left(\dfrac{p_1 v_1}{T_1} + \dfrac{p_2 v_2}{T_2} \right) T$.

$$T = 460 + 72 = 532.$$

Therefore, $P = \dfrac{\left(\dfrac{13 \times 45}{520} + \dfrac{17 \times 60}{540} \right) 532}{60} = 26.723$ lb. per sq.
in. Ans.

(544) One inch corresponds to a pressure of .49 lb.
Therefore, the gauge will show $4.5 \div .49 = 9.18 +$ in.
See Art. 1043.

(545) Sp. Gr. of alcohol $=.8$. Therefore, $16 \times .434 \times .8 =$
pressure exerted by the column of alcohol. $\dfrac{16 \times .434 \times .8}{.49} =$
11.337 in. $=$ height of a column of mercury that will give
the same pressure as 16 ft. of alcohol $=$ number of inches
shown by the gauge. Ans.

(546) (a) $14.7 + 9 = 23.7$ lb. per sq. in. Using
formula 53,

$$v_1 = \frac{pv}{p_1} = \frac{14.7 \times 80}{23.7} = 49.62 \text{ in. } =$$

distance between piston and end of stroke. Since the area
of the piston remains constant, the volume at any point of
the stroke is proportional to the distance passed over by the
piston. Hence, we may use the latter for the former in the
formula. $80 - 49.62 \doteq 30.38$ in. Ans.

(b) Area of piston $= 80' \times .7854$. The volume of air at
point of discharge is $80' \times .7854 \times 49.62$ cu. in. $=$

$$\frac{80' \times .7854 \times 49.62}{1,728} = 144.34 \text{ cu. ft. } \text{ Ans.}$$

(547) Using formula 56, $p W_1 = p_1 W$, or $3.5 \times 14.7 \times 2 =$
$p_1 \times 13$; hence, $p_1 = \dfrac{14.7 \times 3.5 \times 2}{13} = 7.915 +$ lb. per sq. in.
Ans.

(548) $60' - 50' = 10'$. Since the volumes are propor tional to the lengths of the spaces between the piston and the end of the stroke, we may apply formula **62**,

$$\frac{pV}{T} = \frac{p_1 V_1}{T_1}; \text{ or } \frac{14.7 \times 60}{460 + 60} = \frac{p_1 \times 10}{460 + 130}.$$

Therefore, $p_1 = \dfrac{14.7 \times 60 \times 590}{520 \times 10} = 100.07$ lb. per sq. in. Ans.

(549) $T = 127° + 460° = 587°$. Using formula **60**,
$pV = .37052\,T$, or $V = \dfrac{.37052 \times 587}{27} = 8.055$ cu. ft. Ans.

(550) $T = 100° + 460° = 560°$.
Substituting in formula **61**, $pV = .37052\,WT$, or

$$V = \frac{.37052\,WT}{p} = \frac{.37052 \times .5 \times 560}{\dfrac{4,000}{144}} = 3.735 \text{ cu. ft. } \text{Ans.}$$

(551) Use formula **64**. $PV = \left(\dfrac{p_1 v_1}{T_1} + \dfrac{p_2 v_2}{T_2}\right) T.$

$T = 110° + 460° = 570°$; $T_1 = 100° + 460° = 560°$; $T_2 = 130° + 460° = 590°$.

Therefore, $V = \dfrac{\left(\dfrac{90 \times 40}{560} + \dfrac{80 \times 57}{590}\right) 570}{120} = 67.248$ cu. ft. Ans.

(552) The pressure exerted by squeezing the bulb may be found from formula **53**, in which p is 14.7, v, the orig- inal volume = 20 cu. in., and v_1, the new volume, = 5 cu. in.
$p_1 = \dfrac{pv}{v_1} = \dfrac{14.7 \times 20}{5} = 58.8$ lb. The pressure due to the atmosphere must be deducted, since there is an equal pres- sure on the outside which balances it. $58.8 - 14.7 = 44.1$ lb. per sq. in. = pressure due to squeezing the bulb. $3^2 \times .7854 = 7.0686$ sq. in. = area of bottom. $7.0686 \times 44.1 = 311.725$ lb. $7.0686 \times .434 = 3.068$ lb. = pressure due to weight of water. $311.725 + 3.068 = 314.793$ lb. Ans.

(553) Use formula **58**.
$$v_1 = v\left(\frac{460 + t_1}{460 + t}\right) = 4\left(\frac{460 + 115}{460 + 40}\right) = 4.6 \text{ cu. ft. } \text{Ans.}$$

STRENGTH OF MATERIALS.

(554) See Arts. **1094, 1097,** and **1096.**

(555) See Arts. **1102, 1103, 1110,** and **1112.**

(556) See Art. **1105.**

(557) Use formula **67.**

$$E = \frac{Pl}{A\,e}; \text{ therefore, } e = \frac{Pl}{A\,E}.$$

$A = .7854 \times 2^2 ; \; l = 10 \times 12 ; \; P = 40 \times 2,000 ; \; E = 25,000,000.$

$$\text{Therefore, } e = \frac{40 \times 2,000 \times 10 \times 12}{.7854 \times 4 \times 25,000,000} = .122''. \quad \text{Ans.}$$

(558) Using formula **67,**

$$E = \frac{Pl}{A\,e} = \frac{7,000 \times 7\frac{1}{2}}{.7854 \times (\frac{1}{2})^2 \times .009} = 29,708,853.2 \text{ lb. per sq. in.}$$
$$\text{Ans.}$$

(559) Using formula **67,**

$$E = \frac{Pl}{A\,e}, \text{ or } P = \frac{A\,e\,E}{l} = \frac{1\frac{1}{2} \times 2 \times .006 \times 15,000,000}{9 \times 12} =$$

2,500 lb.   Ans.

(560) By formula **67,**

$$E = \frac{Pl}{A\,e}, \text{ or } l = \frac{A\,e\,E}{P} = \frac{.7854 \times 3^2 \times .05 \times 1,500,000}{2,000} = 265.07''.$$
$$\text{Ans.}$$

(561) Using a factor of safety of 4 (see Table 24), formula **65** becomes

$$P = \frac{A\,S_1}{4}, \text{ or } A = \frac{4\,P}{S_1} = \frac{4 \times 6 \times 2{,}000}{55{,}000} = .8727272 \text{ sq. in.}$$

$$d = \sqrt{\frac{A}{.7854}} = \sqrt{\frac{.8727272}{.7854}} = 1.054''. \quad \text{Ans.}$$

(562) From Table 19, the weight of a piece of cast iron 1″ square and 1 ft. long is 3.125 lb.; hence, each foot of length of the bar makes a load of 3.125 lb. per sq. in. The breaking load—that is, the ultimate tensile strength—is 20,000 lb. per sq. in. Hence, the length required to break the bar is $\dfrac{20{,}000}{3.125} = 6{,}400$ ft. Ans.

(563) Let $t =$ the thickness of the bolt head;

$d =$ diameter of bolt.

Area subject to shear $= \pi\,d\,t$.

Area subjected to tension $= \frac{1}{4}\,\pi\,d^2$.

$$S_1 = 55{,}000. \quad S_2 = 50{,}000.$$

Then, in order that the bolt shall be equally strong in both tension and shear, $\pi\,d\,t\,S_2 = \frac{1}{4}\,\pi\,d^2\,S_1$,

$$\text{or } t = \frac{\pi\,d^2\,S_1}{4\,\pi\,d\,S_2} = \frac{d\,S_1}{4\,S_2} = \frac{\frac{3}{4} \times 55{,}000}{4 \times 50{,}000} = .206''. \quad \text{Ans.}$$

(564) Using a factor of safety of 15 for brick, formula **65** gives

$$P = \frac{A\,S_2}{15}.$$

$A = (2\frac{1}{2} \times 3\frac{1}{2})$ sq. ft. $= 30 \times 42 = 1{,}260$ sq. in. ; $S_2 = 2{,}500.$

Therefore, $P = \dfrac{1{,}260 \times 2{,}500}{15} = 210{,}000$ lb. $= 105$ tons. Ans.

(565) The horizontal component of the force P is $P \cos 30° = 3{,}500 \times .866 = 3{,}031$ lb. The area A is $4\,a$, the ultimate shearing strength, S_3, 600 lb., and the factor of safety, 8.

Hence, from formula **65,**

$$P=\frac{A\,S_2}{8}=\frac{4\,a\,S_2}{8}=\frac{a\,S_2}{2}. \quad a=\frac{2\,P}{S_2}=\frac{2\times3,031}{600}=10.1''. \quad \text{Ans.}$$

(566) See Art. **1124.**

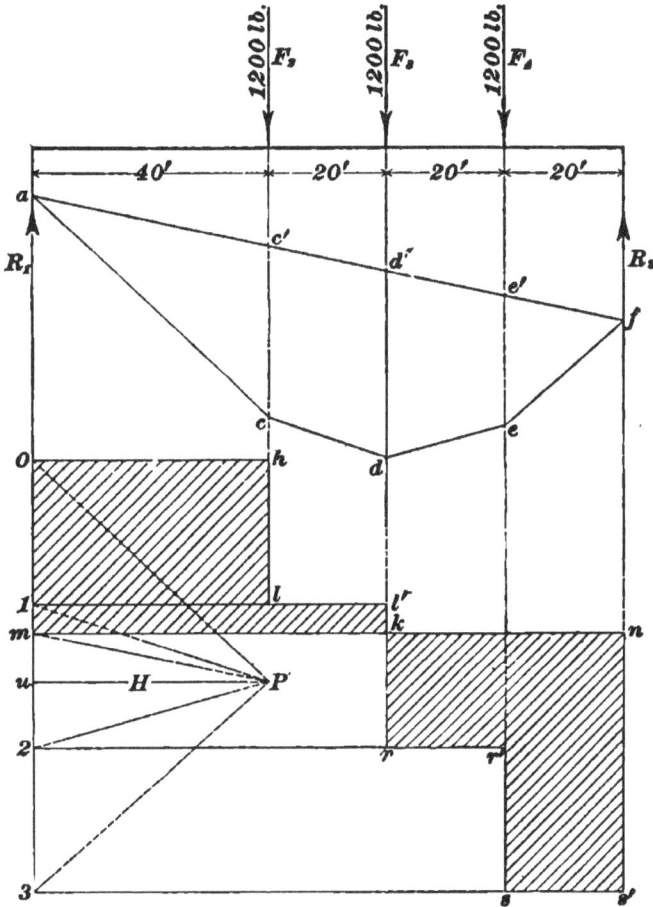

Scale of forces $1''\!\!=\!1600\,lb.$
Scale of distance $1''\!\!=\!32'$
FIG. 46.

(567) Using formula **68,** with the factor of safety of 4,

$$p\,d=\frac{2\,t\,S_1}{4}=\frac{t\,S_1}{2}, \text{ or } t=\frac{2\,p\,d}{S_1}=\frac{2\times120\times48}{55,000}.$$

Since 40% of the plate is removed by the rivet holes, 60% remains, and the actual thickness required is

$$\frac{t}{.60} = \frac{2 \times 120 \times 48}{.60 \times 55,000} = .349''. \quad \text{Ans.}$$

(**568**) Using a factor of safety of 6, in formula **68**,

$$p\,d = \frac{2\,t\,S_1}{6} = \frac{t\,S_1}{3}.$$

Hence, $t = \dfrac{3\,p\,d}{S_1} = \dfrac{3 \times 6 \times 200}{20,000} = .18''. \quad \text{Ans.}$

(**569**) Using formula **71**, with a factor of safety of 10,

$$p = \frac{9,600,000\,t^{2\,.16}}{10\,l\,d} = 960,000\,\frac{t^{2\,.16}}{l\,d}.$$

Hence, $t = \sqrt[2\,.16]{\dfrac{p\,l\,d}{960,000}} = \sqrt[2\,.16]{\dfrac{130 \times 12 \times 12 \times 3}{960,000}} = .272''.$
Ans.

(**570**) From formula **70**,

$$p = \frac{S\,t}{r + t}, \text{ or } t = \frac{p\,r}{S - p} = \frac{2,000 \times \frac{1}{2}}{2,800 - 2,000} = \frac{4,000}{800} = 5''. \quad \text{Ans.}$$

(**571**) See Fig. 46. (*a*) Upon the load line, the loads *0-1*, *1-2*, and *2-3* are laid off equal, respectively, to F_2, F_3, and F_4; the pole *P* is chosen, and the rays drawn in the usual manner; the pole distance $H = 2,000$ lb. The equilibrium polygon is constructed by drawing *a c, c d, d e*, and *e f* parallel to *P 0, P 1, P 2*, and *P 3*, respectively, and finally drawing the closing line *f a* to the starting point *a*. *P m* is drawn parallel to the latter line, dividing the load line into the reactions $m\,0 = R_1$, and $3\,m = R_2$. The shear axis *m n* is drawn through *m*, and the shear diagram *0 h l . . . . s' n m 0* is constructed in the usual manner. To the scale of forces $m\,0 = 1,440$ lb., and $3\,m = 2,160$ lb. To the scale of distances the maximum vertical intercept $y = d'\,d = 31.2$ ft., which, multiplied by H, $= 31.2 \times 2,000 = 62,400$ ft.-lb. $= 748,800$ in.-lb. Ans.

(*b*) The shear at a point 30 ft. from the left support $= 0\,m = 1,440$ lb. Ans.

(*c*) The maximum shear $= n\,s' = -2,160$ lb. Ans.

(572) See Fig. 47. Draw the force polygon *0-1-2-3-4-5-0*
in the usual manner, *0-1* being equal to and parallel to
F_1, *1-2* equal to and parallel to F_2, etc. *0-5* is the resultant.

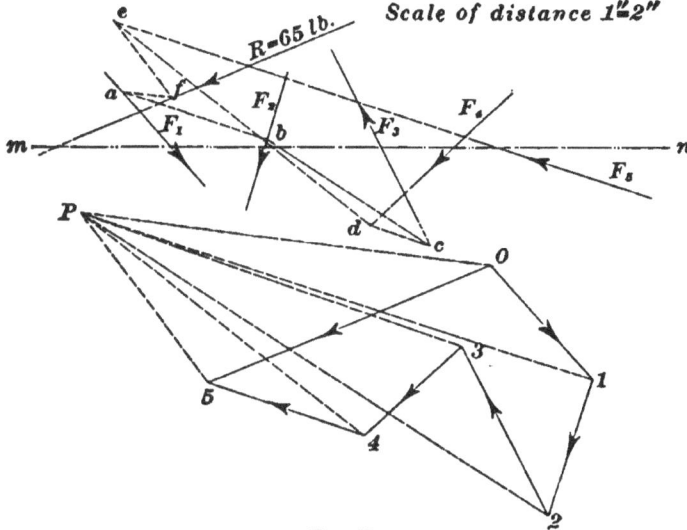

Scale of forces 1=40 lb.
Scale of distance 1"=2"

FIG. 47.

Choose the pole *P*, and draw the rays *P 0*, *P 1*, *P 2*, etc.
Choose any point, *a* on F_1, and draw through it a line
parallel to the ray *P 1*. From the intersection *b* of this line
with F_2, draw a line parallel to *P 2*; from the intersection
c of the latter line with F_3 produced, draw a parallel to
P 3, intersecting F_4 produced in *d*. Finally, through *d*,
draw a line parallel to *P 4*, intersecting F_5 produced in *e*.
Now, through *a* draw a line parallel to *P 0*, and through
e a line parallel to *P 5*; their intersection *f* is a point on the
resultant. Through *f* draw the resultant *R* parallel to
0-5. It will be found by measurement that *R* = 65 lb., that
it makes an angle of $22\frac{1}{2}°$ with *m n*, and intersects it at a
distance of $1\frac{1}{4}'$ from the point of intersection of F_1 and *m n*.

(573) See Fig. 48. The construction is entirely similar
to those given in the text. *0-1*, *1-2*, and *2-3* are laid off to
represent F_1, F_2, and F_3; the pole *P* is chosen and the rays

drawn. Parallel to the rays are drawn the lines of the equilibrium polygon *a b c d g a*. The closing line *g a* is found to be parallel to *P 1*. Consequently, *0-1* is the left reaction and *1-3* the right reaction, the former being 6 tons

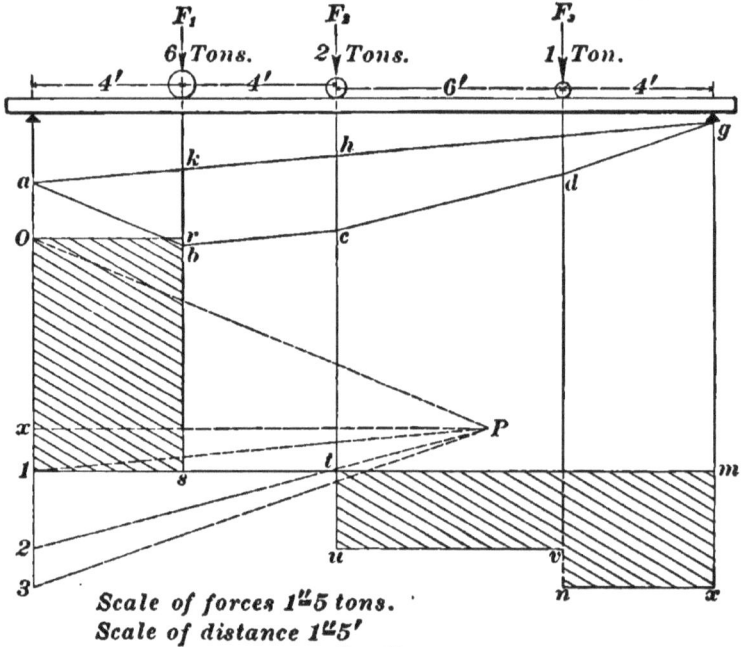

F_1 F_2 F_3

6 *Tons.*   2 *Tons.*   1 *Ton.*

Scale of forces 1″5 tons.
Scale of distance 1″5′

FIG. 48.

and the latter 3 tons. The shear diagram is drawn in the usual manner; it has the peculiarity of being zero between F_1 and F_2.

(574) The maximum moment occurs when the shear line crosses the shear axis. In the present case the shear line and shear axis coincide with *s t*, between F_1 and F_2; hence, the bending moment is the same (and maximum) at F_1 and F_2, and at all points between. This is seen to be true from the diagram, since *k h* and *b c* are parallel. Ans.

(*b*) By measurement, the moment is found to be 24×12= 288 inch-tons. Ans.

(*c*) 288 × 2,000 = 576,000 inch-pounds. Ans.

(575)　See Arts. **1133** to **1137.**

(576)　See Fig. 49.　The force polygon *0-1-2-3-4-0* is drawn as in Fig. 47, *0-4* being the resultant.　The equilibrium polygon *a b c d g a* is then drawn, the point *g* lying on the resultant.　The resultant *R* is drawn through *g*, parallel to and equal to *0-4*.　A line is drawn through *C*, parallel to *R*. Through *g* the lines *g e* and *g f* are drawn parallel, respectively, to *P0* and *P4*, and intersecting the parallel to *R*, through *C* in *e* and *f* ; then, *ef* is the intercept, and *Pu*, perpendicular to *0-4*, is the pole distance.　$Pu = 33$ lb. ; $ef = 1.32''$.　Hence, the resultant moment is $33 \times 1.32 = 43.6$ in.-lb.　Ans.

Scale of forces 1″50 lb.
Scale of distance 1″2″

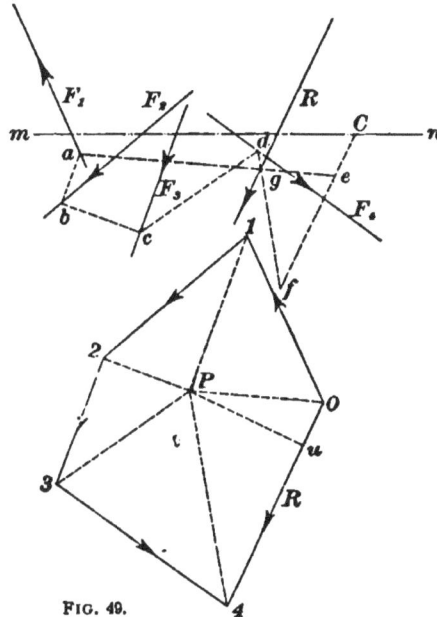

FIG. 49.

(577)　The maximum bending moment, $M = W\dfrac{l_1\, l_2}{l}$ (see Fig. 6 of table of Bending Moments) $= 4 \times 2{,}000 \times \dfrac{14 \times 8}{22} =$ $40{,}727\frac{3}{11}$ ft.-lb. $= 488{,}727$ in.-lb.　Then, according to formula **74,**

$$\frac{S_4 I}{f c} = 488{,}727.$$

$$\frac{I}{c} = \frac{488{,}727 f}{S_4} = \frac{488{,}727 \times 8}{9{,}000} = 434.424.$$

But, $\dfrac{I}{c} = \dfrac{\frac{1}{12}\,b\,d^3}{\frac{1}{2}\,d} = \dfrac{1}{6}\,b\,d^2$, and, according to the conditions

of the problem, $b = \dfrac{1}{2}\,d$.

Therefore, $\dfrac{I}{c} = \dfrac{1}{6}\,b\,d^2 = \dfrac{1}{12}d^3 = 434.424$.

$$d^3 = 5,213.088.$$
$$\left.\begin{array}{l} d = 17\tfrac{1}{3}''. \\ b = 8\tfrac{2}{3}''. \end{array}\right\}\ \text{Ans.}$$

(578) The beam, with the moment and shear diagrams, is shown in Fig. 50. On the line, through the left reaction,

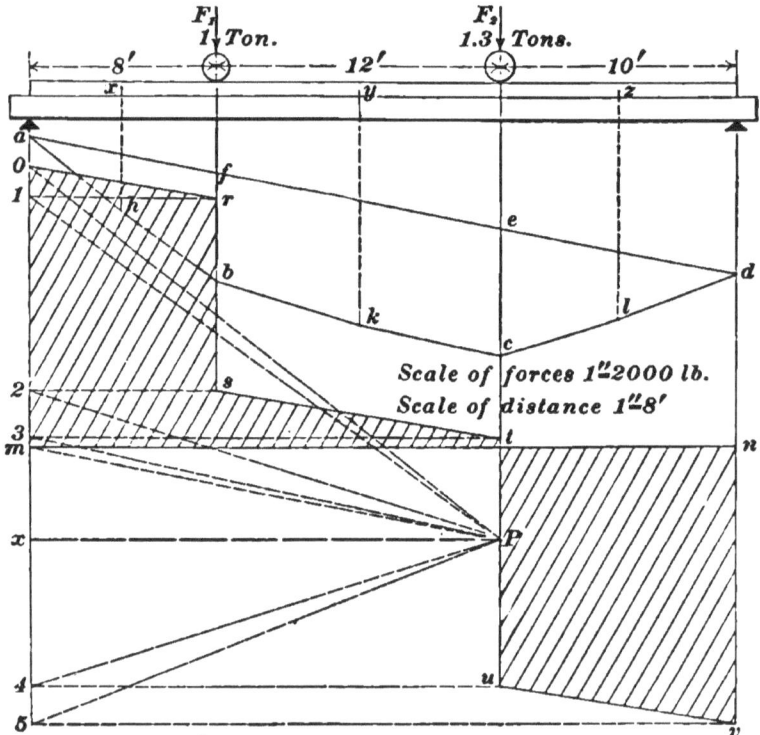

FIG. 50.

are laid off the loads in order. Thus, $0\text{-}1 = 40 \times 8 = 320$ lb., is the uniform load between the left support and F_1; $1\text{-}2$ is

$F_1 = 2,000$ lb. ; $2\text{-}3 = 40 \times 12 = 480$ lb., is the uniform load between F_1 and F_2; $3\text{-}4 = 2,000 \times 1.3 = 2,600$ lb., is F_2, and $4\text{-}5 = 40 \times 10 = 400$ lb., is the uniform load between F_2 and the right support. The pole P is chosen and the rays drawn. Since the uniform load is very small compared with F_1 and F_2, it will be sufficiently accurate to consider the three portions of it concentrated at their respective centers of gravity x, y, and z. Drawing the equilibrium polygon parallel to the rays, we obtain the moment diagram $a\ h\ b\ k\ c\ l\ d\ a$. From P, drawing $P\,m$ parallel to the closing line $a\ d$, we obtain the reactions $0\ m$ and $m\ 5$ equal, respectively, to 2,930 and 2,870 lb. Ans. The shear axis $m\ x$, and the shear diagram $0\ r\ s\ t\ u\ v\ n\ m$, are drawn in the usual manner. The greatest shear is $0\ m$, 2,930 lb. The shear line cuts the shear axis under F_2. Hence, the maximum moment is under F_2. By measurement, $e\ c$ is 64″, and $P\ x$ is 5,000 lb. ; hence, the maximum bending moment is $64 \times 5,000 = 320,000$ in.-lb. Ans.

(**579**) From the table of Bending Moments, the greatest bending moment of such a beam is $\dfrac{w\,l^2}{8}$, or, in this case, $\dfrac{w \times 240^2}{8}$.

By formula **74**,

$$M = \frac{w \times 240^2}{8} = \frac{S_t I}{fc} = \frac{45,000}{4} \times \frac{280}{12 \div 2}.$$

Therefore, $w = \dfrac{45,000 \times 280 \times 8}{240 \times 240 \times 4 \times 6} = 72.92$ lb. per inch of length $= 72.92 \times 12 = 875$ lb. per foot of length. Ans.

(**580**) From the table of Bending Moments, the maximum bending moment is

$$\frac{W l}{4} = \frac{W \times 96}{4} = 24\ W.$$

From formula **74**,

$$M = 24\ W = \frac{S_t I}{fc}.$$

$I = \dfrac{\pi}{64}(d^4 - d_1^4) = 56.945\,;\ c = \dfrac{1}{2}\,d = \dfrac{6\frac{1}{2}}{2} = 3\frac{1}{4}\,;\ S_t = 38,000\,;\ f = 6.$

Hence, $24\ W = \dfrac{38,000 \times 56.945}{6 \times 3.25}$.

$$W = \dfrac{38,000 \times 56.945}{24 \times 6 \times 3.25} = 4,024\ \text{lb.}\quad\text{Ans.}$$

(581) (a) From the table of Bending Moments,

$$M = \dfrac{w\,l^2}{8} = \dfrac{w \times 192^2}{8}.$$

From formula **74**,

$$M = \dfrac{w \times 192^2}{8} = \dfrac{S_i\,I}{f\,c}.$$

$$S_i = 7,200;\ f = 8;\ I = \dfrac{1}{12}b\,d^2 = \dfrac{2,000}{12};\ c = \dfrac{1}{2}d = 5.$$

Then, $\quad\dfrac{w \times 192^2}{8} = \dfrac{7,200}{8} \times \dfrac{2,000}{12 \times 5}.$

$$w = \dfrac{7,200 \times 2,000 \times 8}{8 \times 12 \times 5 \times 192 \times 192} =$$

6.51 lb. per in. $= 6.51 \times 12 = 78.12$ lb. per ft.   Ans.

(b) $I = \dfrac{1}{12}b\,d^2 = \dfrac{10 \times 2^2}{12} = \dfrac{80}{12}.\quad c = 1''.$

$$\dfrac{w \times 192^2}{8} = \dfrac{7,200}{8} \times \dfrac{80}{12 \times 1}.$$

$$w = \dfrac{7,200 \times 80 \times 8}{8 \times 12 \times 192 \times 192} = 1.3\ \text{lb. per in.}$$

$$= 1.3 \times 12 = 15.6\ \text{lb. per ft.}\quad\text{Ans.}$$

(582) (a) From the table of Bending Moments, the deflection of a beam uniformly loaded is $\dfrac{5}{384}\dfrac{Wl^3}{EI}$. In Example 579, $W = 874 \times 20 = 17,480$ lb.; $l = 240''$, $E = 25,000,000$, and $I = 280$.

Hence, deflection $s = \dfrac{5 \times 17,480 \times 240^3}{384 \times 25,000,000 \times 280} = .45$ in.   Ans.

(b) From the table of Bending Moments, $s = \dfrac{1}{48}\dfrac{Wl^3}{EI}$.

In Example 580, $W = 4,624$ lb.; $l = 96$ in.; $E = 15,000,000$, and $I = 56.945$.

Hence, $s = \dfrac{4,624 \times 96^3}{48 \times 15,000,000 \times 56.945} = .1''$, nearly. Ans.

(c) $s = \dfrac{5}{384} \dfrac{Wl^3}{EI}$. In Example 581 (a), $W = 78.12 \times 16$; $l = 192$; $E = 1,500,000$, and $I = \dfrac{2,000}{12}$.

Hence, $s = \dfrac{5 \times 78.12 \times 16 \times 192^3}{384 \times 1,500,000 \times \frac{2000}{12}} = .461''$. Ans.

(583) Area of piston $= \dfrac{1}{4}\pi d^2 = \dfrac{1}{4}\pi \times 14^2$.

$W =$ pressure on piston $= \dfrac{1}{4}\pi \times 14^2 \times 80$.

From the table of Bending Moments, the maximum bending moment for a cantilever uniformly loaded is

$\dfrac{wl^2}{2} = \dfrac{Wl}{2} = \dfrac{\frac{1}{4}\pi \times 14^2 \times 80 \times 4}{2} = \dfrac{S_4}{f}\dfrac{I}{c}$. See formula **74.**

$\dfrac{S_4}{f} = \dfrac{45,000}{10} = 4,500$. $\dfrac{I}{c} = \dfrac{\frac{1}{64}\pi d^4}{\frac{1}{2}d} = \dfrac{1}{32}\pi d^3$.

Hence, $\dfrac{\frac{1}{4}\pi \times 14^2 \times 80 \times 4}{2} = \dfrac{4,500\,\pi d^3}{32}$,

or $d^3 = \dfrac{14^2 \times 80 \times 4 \times 32}{4 \times 2 \times 4,500} = 55.75$.

$d = \sqrt[3]{55.75} = 3.82''$. Ans.

(584) Substituting in formula **76,** $S_2 = 90,000$; $A = 6^2 \times .7854$; $f = 6$; $l = 14 \times 12 = 168$; $g = 5,000$; $I = \dfrac{\pi}{64} \times 6^4$, we obtain

$W = \dfrac{S_2 A}{f\left(1 + \dfrac{Al^2}{gI}\right)} = \dfrac{90,000 \times 6^2 \times .7854}{6\left(1 + \dfrac{6^2 \times .7854 \times 168^2}{5,000 \times \frac{3.1416 \times 6^4}{64}}\right)} = 120,872$ lb. Ans.

(585) For timber, $S_2 = 8,000$ and $f = 8$; hence, $\dfrac{S_2}{f} = \dfrac{8,000}{8} = 1,000$.

Substituting in formula **65**,

$$P = A \frac{S_2}{f} = 1,000\,A.$$

$$A = \frac{P}{1,000} = \frac{7 \times 2,000}{1,000} = 14 \text{ sq. in., necessary area of a}$$

short column to support the given load. Since the column is quite long, assume it to be 6″ square. Then $A = 36$, and $I = \frac{1}{12}\,b^4 = \frac{6^4}{12} = 108.$

Formula **76** gives

$$W = \frac{S_2 A}{f\left(1 + \frac{A\,l^2}{g\,I}\right)}, \text{ or } \frac{S_2}{f} = \frac{W}{A}\left(1 + \frac{A\,l^2}{g\,I}\right).$$

$l = 30 \times 12 = 360$, and $g = 3,000$.

$$\frac{S_2}{f} = \frac{14,000}{36}\left(1 + \frac{36 \times 360^2}{3,000 \times 108}\right) = 5,990, \text{ nearly.}$$

Since this value is much too large, the column must be made larger. Trying 9″ square, $A = 81$, $I = 546\frac{3}{4}$.

Then, $\frac{S_2}{f} = \frac{14,000}{81}\left(1 + \frac{81 \times 360 \times 360}{3,000 \times 546\frac{3}{4}}\right) = 1,279.$

This value of $\frac{S_2}{f}$ is much nearer the required value, 1,000.

Trying 10″ square, $A = 100$, $I = \frac{10,000}{12} = 833\frac{1}{3}.$

$$\frac{S_2}{f} = \frac{14,000}{100}\left(1 + \frac{100 \times 360 \times 360}{3,000 \times 833\frac{1}{3}}\right) = 866, \text{ nearly.}$$

Since this value of $\frac{S_2}{f}$ is less than 1,000, the column is a little too large; hence, it is between 9 and 10 inches square. $9\frac{5}{8}″$ will give 997.4 lb. as the value of $\frac{S_2}{f}$; hence, the column should be $9\frac{5}{8}″$ square.

This problem may be more readily solved by formula **77**, which gives

$$c = \sqrt{\frac{7 \times 2000 \times 8}{2 \times 800} + \sqrt{\frac{7 \times 2000 \times 8}{8000}\left(\frac{7 \times 2000 \times 8}{4 \times 8000} + \frac{12 \times 360^2}{3000}\right)}} =$$

$$\sqrt{7 + \sqrt{14\,(3.5 + 518.4)}} = \sqrt{92.479} = 9.61' = 9\frac{5}{8}', \text{ nearly.}$$

(586) Here $W = 21,000$; $f = 10$; $S_1 = 150,000$; $g = 6,250$; $l = 7.5 \times 12 = 90''$. For using formula **78**, we have

$$\frac{.3183 \; Wf}{S_1} = \frac{.3183 \times 21,000 \times 10}{150,000} = .4456.$$

$$\frac{16 \, l^2}{g} = \frac{16 \times 8100}{6250} = 20.7360.$$

Therefore,

$$d = 1.4142 \sqrt{.4456 + \sqrt{.4456 \, (.4456 + 20.7360)}} =$$
$$1.4142 \sqrt{.4456 + 3.0722} = 2.65'', \text{ or say } 2\tfrac{5}{8}''.$$

(587) For this case, $A = 3.1416$ sq. in. ; $l = 4 \times 12 = 48''$; $S_1 = 55,000$; $f = 10$; $I = .7854$; $g = 20,250$. Substituting these values in formula **76**,

$$W = \frac{S_1 A}{f\left(1 + \dfrac{A \, l^2}{g \, I}\right)} = \frac{55,000 \times 3.1416}{10\left(1 + \dfrac{3.1416 \times 48^2}{20,250 \times .7854}\right)} =$$
$$\frac{5,500 \times 3.1416}{1.4551}.$$

Steam pressure $= 60$ lb. per sq. in.

Then, area of piston $= .7854 \, d' = \dfrac{W}{60} = \dfrac{5,500 \times 3.1416}{1.4551 \times 60}.$

Hence, $d^2 = \dfrac{5,500 \times 3.1416}{.7854 \times 1.4551 \times 60} = 252$, nearly,

and $d = \sqrt{252} = 15\tfrac{7}{8}''$, nearly. Ans.

(588) (*a*) The strength of a beam varies directly as the width and square of the depth and inversely as the length. Hence, the ratio between the loads is

$$\frac{6 \times 8^2}{10} : \frac{4 \times 12^2}{16} = 16 : 15, \text{ or } 1\tfrac{1}{15}. \quad \text{Ans.}$$

(*b*) The deflections vary directly as the cube of the lengths, and inversely as the breadths and cubes of the depths. Hence, the ratio between the deflections is

$$\frac{10^3}{6 \times 8^3} : \frac{16^3}{4 \times 12^3} = .549. \quad \text{Ans.}$$

(589) Substituting the value of c_1, from Table 27. in formula **80,** we obtain

(a) $d = c_1 \sqrt[4]{\dfrac{H}{N}} = 4.92 \sqrt[4]{\dfrac{40}{120}} = 3.739''$.   Ans.

(b) $d = c_1 \sqrt[4]{\dfrac{H}{N}} = 4.92 \sqrt[4]{\dfrac{80}{100}} = 4.65''$.   Ans.

(590) Using formula **80,**

$$d = 5.59 \sqrt[4]{\dfrac{4,000}{50}} = 14.06''.$$

Since this result is greater than 13.6", formula **81** must be used, in which

$$d = k_1 \sqrt[3]{\dfrac{H}{N}} = 3.3 \sqrt[3]{\dfrac{4,000}{50}} = 14.22''.$$ Ans.

(591) From formula **80,**

$$d = c_1 \sqrt[4]{\dfrac{H}{N}}, \text{ or } H = \dfrac{d^4 N}{c_1^4}. \quad c_1 = 4.11. \quad \text{(Table 27.)}$$

Hence, $H = \dfrac{4^4 \times 80}{4.11^4} = 71.775$ H. P.   Ans.

(592) Using formula **83,**

$$H = q_1 N \left(\dfrac{d_1^4 - d_2^4}{d_1}\right) = .0212 \times 100 \left(\dfrac{(7\frac{1}{2})^4 - 5^4}{7\frac{1}{2}}\right) = 717.7 \text{ H.P. Ans}$$

(.0212 is the value of q_1 from Table 28.)

(593) (a) Using formula **84,**
$$P = 100\,C^2 = 100 \times 8^2 = 6,400 \text{ lb.}$$ Ans.

(b) Using formula **85,**
$$P = 600\,C^2, \text{ or } C^2 = \dfrac{P}{600} = \dfrac{6,000}{600} = 10.$$
$$C = \sqrt{10} = 3.162''.$$
$$d = \dfrac{1}{3}C = 1.054''.$$ Ans.

(c) Using formula **86,**
$$P = 1,000\,C^2; \quad C^2 = \dfrac{P}{1,000} = \dfrac{6\frac{2}{3} \times 2,000}{1,000} = 13\frac{1}{3}.$$
$$C = \sqrt{13\frac{1}{3}} = 3.651''.$$ Ans.

(594) (*a*) Using formula **87**,

$$P = 12,000 \, d^2 = 12,000 \times \left(\frac{7}{8}\right)^2 = 9,187.5 \text{ lb.} \quad \text{Ans.}$$

(*b*) Formula **88** gives $P = 18,000 \, d^2$.

Therefore, $d = \sqrt{\dfrac{P}{18,000}} = \sqrt{\dfrac{8,000}{18,000}} = \sqrt{\dfrac{4}{9}} = .667''$. Ans.

(595) The deflection is, by formula **75**,

$s = a \dfrac{W l^3}{E I} = \dfrac{1}{192} \dfrac{W l^3}{E I}$, the coefficient being found from the table of Bending Moments.

Transposing, $W = \dfrac{192 \, s \, E \, I}{l^3}$; $l = 120$; $E = 30,000,000$; $I = .7854$; $s = \dfrac{1}{8}$.

Then, $W = \dfrac{192 \times 30,000,000 \times .7854}{8 \times 120^3} = 327.25$ lb. Ans.

(596) (*a*) The maximum bending moment is, according to the table of Bending Moments, $\dfrac{W l}{4} = \dfrac{6,000 \times 60}{4} = 90,000$ inch-pounds.

By formula **74**,

$$M = 90,000 = \frac{S_t}{f} \frac{I}{c}.$$

$S_t = 120,000$; $f = 10$. $\dfrac{I}{c} = \dfrac{\dfrac{\pi d^4}{64}}{\frac{1}{2} d} = \dfrac{\pi d^3}{32}$.

Hence, $\qquad 90,000 = \dfrac{120,000}{10} \dfrac{\pi d^3}{32}$,

or $d = \sqrt[3]{\dfrac{90,000 \times 10 \times 32}{120,000 \times 3.1416}} = 4.244'' = 4\frac{1}{4}''$, nearly.

(*b*) Using formula **80**,

$$d = c_t \sqrt[3]{\frac{H}{N}} = 4.7 \sqrt[3]{\frac{75}{80}} = 4\frac{5}{8}'', \text{ nearly.} \quad \text{Ans.}$$

(597) (*a*) The graphic solution is shown in Fig. 51.
On the vertical through the support *0-1* is laid off equal to
the uniform load between the support and F_1; *1-2* is laid off

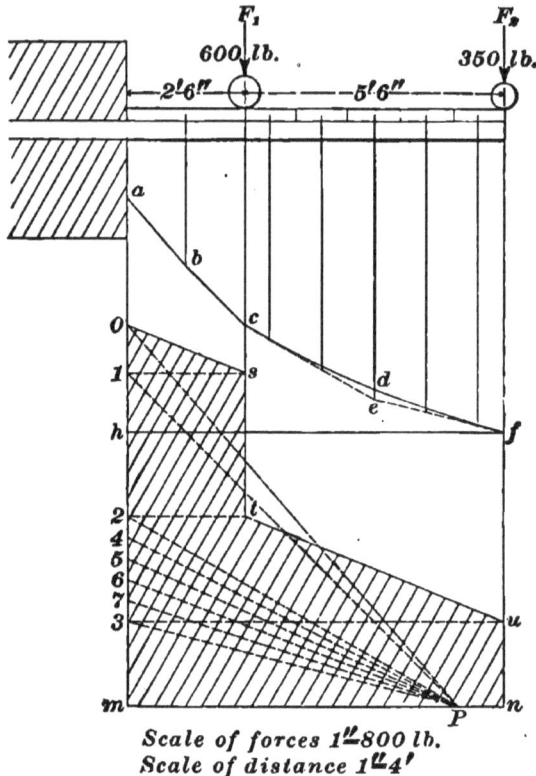

Scale of forces 1″=800 lb.
Scale of distance 1″=4′

FIG. 51.

to represent F_1. *2-3* represents to the same scale the remainder of the uniform load, and *3 m* represents F_2. The
pole *P* is chosen and the rays drawn. The polygon *a b c e f h*
is then drawn, the sides being parallel, respectively, to the
corresponding rays. If the uniform load between F_1 and F_2
be considered as concentrated at its center of gravity, the
polygon will follow the broken line *c e f*. It will be better
in this case to divide the uniform load into several parts,
2-4, 4-5, 5-6, etc., thus obtaining the line of the polygon

$c\,d\,f$. To draw the shear diagram, project the point 1 across the vertical through F_1, and draw $O\,s$. Next project the point 2 across to t, and 3 across to u, and draw $t\,u$. $O\,s\,t\,u\,n\,m$ is the shear diagram. The maximum moment is seen to be at the support, and is equal to $a\,h \times P\,m$. To the scale of distances, $a\,h = 58.8$ in., while $P\,m = H = 1,400$ lb. to the scale of forces. Hence, the maximum bending moment is $58.8 \times 1,400 = 82,320$ in.-lb.   Ans.

(b) From formula **74**,

$$M = \frac{S_t}{f}\frac{I}{c} = 82,320. \quad S_t = 12,500;\; f = 8.$$

$$\text{Therefore, } \frac{I}{c} = \frac{82,320 \times 8}{12,500} = 52.68.$$

$$\text{But, } \frac{I}{c} = \frac{\frac{1}{12}b\,d^3}{\frac{1}{2}d} = \frac{b\,d^2}{6}, \text{ and } d = 2\tfrac{1}{2}b, \text{ or } b = \frac{2\,d}{5}.$$

$$\text{Hence, } \frac{I}{c} = \frac{b\,d^2}{6} = \frac{d^3}{15} = 52.68. \quad d^3 = 52.68 \times 15 = 790.2.$$

$$d = \sqrt[3]{790.2} = 9.245". \quad b = \frac{2\,d}{5} = 3.7", \text{ nearly.   Ans.}$$

(**598**) Referring to the table of Moments of Inertia,

$$I = \frac{(b\,d^2 - b_1\,d_1{}^2)^2 - 4\,b\,d\,b_1\,d_1(d - d_1)^2}{12\,(b\,d - b_1\,d_1)} =$$

$$\frac{[8 \times 10^2 - 6 \times (8\tfrac{1}{2})^2]^2 - 4 \times 8 \times 10 \times 6 \times 8\tfrac{1}{2}\,(10 - 8\tfrac{1}{2})^2}{12\,(8 \times 10 - 6 \times 8\tfrac{1}{2})} =$$

$$280.466.$$

$$c = \frac{d}{2} + \frac{b_1\,d_1}{2}\left(\frac{d - d_1}{b\,d - b_1\,d_1}\right) =$$

$$\frac{10}{2} + \frac{6 \times 8\tfrac{1}{2}}{2}\left(\frac{10 - 8\tfrac{1}{2}}{8 \times 10 - 6 \times 8\tfrac{1}{2}}\right) = 6.319.$$

(a) From the table of Bending Moments, the maximum bending moment is $\dfrac{W\,l}{4}$.

$$S_t = 120,000;\; f = 7;\; l = 35 \times 12 = 420 \text{ in.}$$

$$\text{Using formula } \textbf{74}, \; M = \frac{W\,l}{4} = \frac{S_t\,I}{f\,c}, \text{ or}$$

$$W = \frac{4\,S_t\,I}{l\,f\,c} = \frac{4 \times 120,000 \times 280.466}{420 \times 7 \times 6.319} = 7.246 \text{ lb.   Ans.}$$

(*b*) In this case $f = 5$, and the maximum bending moment is $\dfrac{w\,l^2}{8}$. Hence, from formula **74**,

$$M = \frac{w\,l^2}{8} = \frac{S_t\,I}{fc}, \text{ or } w = \frac{8\,S_t\,I}{l^2\,fc}.$$

Therefore, $W = w\,l = \dfrac{8\,S_t\,I}{l\,fc} = \dfrac{8 \times 120{,}000 \times 280.466}{420 \times 5 \times 6{,}319} =$ 20,290 lb. Ans.

(**599**) (*a*) According to formula **72**,

$$I = A\,r^2, \text{ or } r = \sqrt{\frac{I}{A}} = \sqrt{\frac{72}{24}} = \sqrt{3} = 1.732. \text{ Ans.}$$

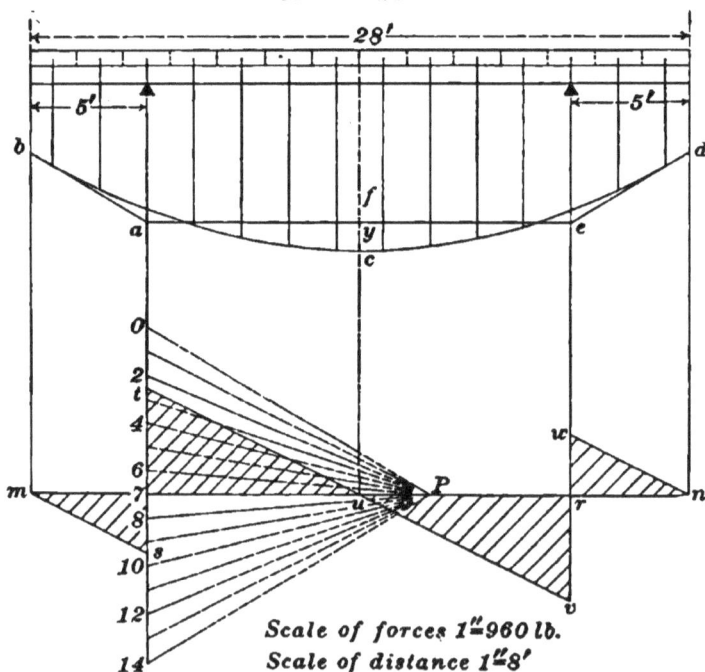

Scale of forces 1"=960 lb.
Scale of distance 1"=8'

FIG. 52.

(*b*) From the table of Moments of Inertia, $I = \dfrac{1}{12}\,b\,d^3 = 72$; $A = b\,d = 24$. Dividing, $\dfrac{\frac{1}{12}\,b\,d^3}{b\,d} = \dfrac{72}{24}$, or $\dfrac{1}{12}\,d^2 = 3$. $d^2 = 36$; $d = 6'$ and $b = 4'$. Ans.

(c) As above, $r = \sqrt{\dfrac{I}{A}} = \sqrt{\dfrac{\frac{\pi d^4}{64}}{\frac{\pi d^2}{4}}} = \sqrt{\dfrac{d^2}{16}} = \dfrac{d}{4}$. Ans.

(**600**) Using formula **69**, $pd = 4tS$, we have $t = \dfrac{pd}{4S}$. Using a factor of safety of 6,

$$pd = \frac{4tS}{6}, \text{ or } t = \frac{6pd}{4S} = \frac{6 \times 100 \times 8}{4 \times 20,000} = .06''. \text{ Ans.}$$

(**601**) The graphic solution is shown in Fig. 52. The uniform load is divided into 14 equal parts, and lines drawn through the center of gravity of each part. These loads are laid off on the line through the left reaction, the pole P chosen, and the rays drawn. The polygon $b\,c\,d\,e\,f\,a$ is then drawn in the usual manner. The shear diagram is drawn as shown. The maximum shear is either $t\,7$ or $r\,v = 540$ lb. The maximum moment is shown by the polygon to be at $f\,c$ vertically above the point u, where the shear line crosses the shear axis. The pole distance $P7$ is 1,440 lb. to the scale of forces, and the intercept $f\,c$ is 14 inches to the scale of distances. Hence, the bending moment is 20,160 in.-lb.

(**602**) From formula **74**,

$$M = \frac{S_t}{f}\frac{I}{c} = 20,160. \quad S_t = 9,000; \ f = 8.$$

$$\text{Then, } \frac{I}{c} = \frac{20,160 \times 8}{9,000} = 17.92.$$

$$\text{But, } \frac{I}{c} = \frac{\frac{1}{12} b d^3}{\frac{1}{2}d} = \frac{1}{6} b d^2 \text{ for a rectangle.}$$

$$\text{Hence, } \frac{1}{6} b d^2 = 17.92, \text{ or } b d^2 = 107.52.$$

Any number of beams will fulfil this condition.

$$\text{Assuming } d = 6'', \ b = \frac{107.52}{36} = 3'', \text{ nearly.}$$

$$\text{Assuming } d = 5'', \ b = \frac{107.52}{25} = 4.3''.$$

(**603**) Using the factor of safety of 10, in formula **71**,

$$p = \frac{9,600,000}{10}\frac{t^{2.1}}{ld} = \frac{960,000 \times .2^{2.1}}{108 \times 2.5} = 106.45 \text{ lb.} \quad \text{Ans.}$$

(604) Using formula **87,**

$$P = 12,000\,d^3, \text{ or } d = \sqrt{\frac{P}{12,000}} = \sqrt{\frac{5 \times 2,000}{12,000}} = .913''. \text{ Ans.}$$

(605) The radius r of the gear-wheel is $24''$. Using formula **80,** $d = c \sqrt[3]{P\,r} = .297 \sqrt[3]{350 \times 24} = 2.84''$. Ans.

(606) Area of cylinder $= .7854 \times 12^2 = 113.1$ sq. in.
Total pressure on the head $= 113.1 \times 90 = 10,179$ lb.

Pressures on each bolt $= \dfrac{10,179}{10} = 1,017.9$ lb.

Using formula **65,**

$$P = A\,S, \text{ or } A = \frac{P}{S} = \frac{1,017.9}{2,000} = .5089 \text{ sq. in., area of bolt.}$$

Diameter of bolt $= \sqrt{\dfrac{.5089}{.7854}} = .8''$, nearly. Ans.

(607) (*a*) The graphic solution is clearly shown in Fig. 53. On the vertical through F_1, the equal loads F_1 and F_2

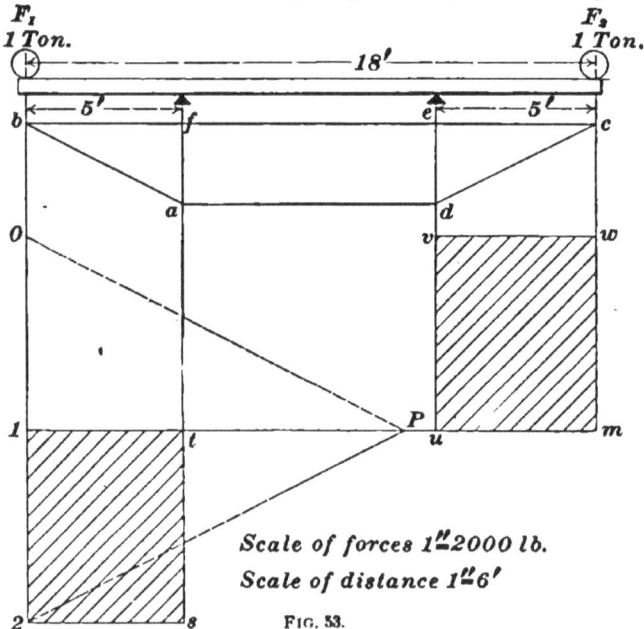

Scale of forces $1'' = 2000$ lb.
Scale of distance $1'' = 6'$

FIG. 53.

are laid off to scale, *0-1* representing F_1 and *1-2* representing

F_2. Choose the pole P, and draw the rays P 0, P 1, P 2. Draw a b between the left support and F_1 parallel to P 0; b c between F_1 and F_2 parallel to P 1, and c d parallel to P 2, between F_2 and the right support. Through P draw a line parallel to the closing line a d. 0-$1 = 1$-2; hence, the reactions of the supports are equal, and are each equal to 1 ton. The shear between the left reaction and F_1 is negative, and equal to $F_1 = 1$ ton. Between the left and the right support it is 0, and between the latter and F_2 it is positive and equal to 1 ton. The bending moment is constant and a maximum between the supports. To the scale of forces P $1 = 2$ tons $= 4,000$ lb., and to the scale of distances a $f = 30$ in. Hence, the maximum bending is $4,000 \times 30 = 120,000$ in.-lb. Ans.

(*b*) Using formula **74,**

$$M = \frac{S_4}{f} \frac{I}{c} = 120,000. \quad S_4 = 38,000; \quad f = 6.$$

Then, $\dfrac{I}{c} = \dfrac{120,000 \times 6}{38,000} = \dfrac{360}{19} = 19$, nearly.

But, $\dfrac{I}{c} = \dfrac{\frac{\pi d^4}{64}}{\frac{d}{2}} = \dfrac{\pi d^3}{32}.$

Hence, $\dfrac{\pi d^3}{32} = 19$, or $d^3 = \dfrac{32 \times 19}{3.1416}.$

$$d = \sqrt[3]{\frac{32 \times 19}{3.1416}} = 5.784''. \quad \text{Ans.}$$

(608) Since the deflections are directly as the cubes of the lengths, and inversely as the breadths and the cubes of the depths, their ratio in this case is

$$\frac{18^3}{2 \times 6^3} : \frac{12^3}{3 \times 8^3} \text{ or } \frac{27}{2} : \frac{9}{8} = 12.$$

That is, the first beam deflects 12 times as much as the second. Hence, the required deflection of the second beam is $.3 \div 12 = .025'$. Ans.

(609) The key has a shearing stress exerted on two sections; hence, each section must withstand a stress of $\dfrac{20,000}{2} = 10,000$ pounds.

Using formula **65,** with a factor of safety of 10,

$$P = \frac{A\,S_s}{10}, \text{ or } A = \frac{10\,P}{S_s} = \frac{10 \times 10,000}{50,000} = 2 \text{ sq. in.}$$

Let $b =$ width of key;

$t =$ thickness.

Then, $b\,t = A = 2$ sq. in. But, from the conditions of the problem,

$$t = \frac{1}{4}b.$$

Hence, $b\,t = \dfrac{1}{4}b^2 = 2;\ b^2 = 8;\ b = 2.828''.$

$$t = \frac{2.828}{4} = .707''. \qquad \Bigg\} \text{ Ans.}$$

(610) From formula **75,** the deflection $S = a\,\dfrac{W\,l^3}{E\,I}$, and, from the table of Bending Moments, the coefficient a for the beam in question is $\dfrac{1}{48}$.

$W = 30$ tons $= 60,000$ lb.; $l = 54$ inches; $E = 30,000,000$; $I = \dfrac{\pi\,d^4}{64}$.

Hence, $S = \dfrac{1 \times 60,000 \times 54^3}{48 \times 30,000,000 \times \dfrac{3.1416 \times 12^4}{64}} = .0064$ in.　Ans.

(611) (*a*) The circumference of a 7-strand rope is 3 times the diameter; hence, $C = 1\frac{1}{4} \times 3 = 3\frac{3}{4}''.$

Using formula **86,** $P = 1,000\ C^2 = 1,000 \times (3\frac{3}{4})^2 = 14,062.5$ lb.　Ans.

(*b*) Using formula **84,**

$$P = 100\,C^2, \text{ or } C = \sqrt{\frac{P}{100}} = \sqrt{\frac{1\frac{1}{4} \times 2,000}{100}} = 5.92''.　\text{ Ans.}$$

(612)   Using formula **70,**

$$p = \frac{S_1 t}{r+t} = \frac{120,000}{\frac{8}{2}+6} = 12,000 \text{ lb.}\quad \text{Ans.}$$

(613)   The construction of the diagram of bending

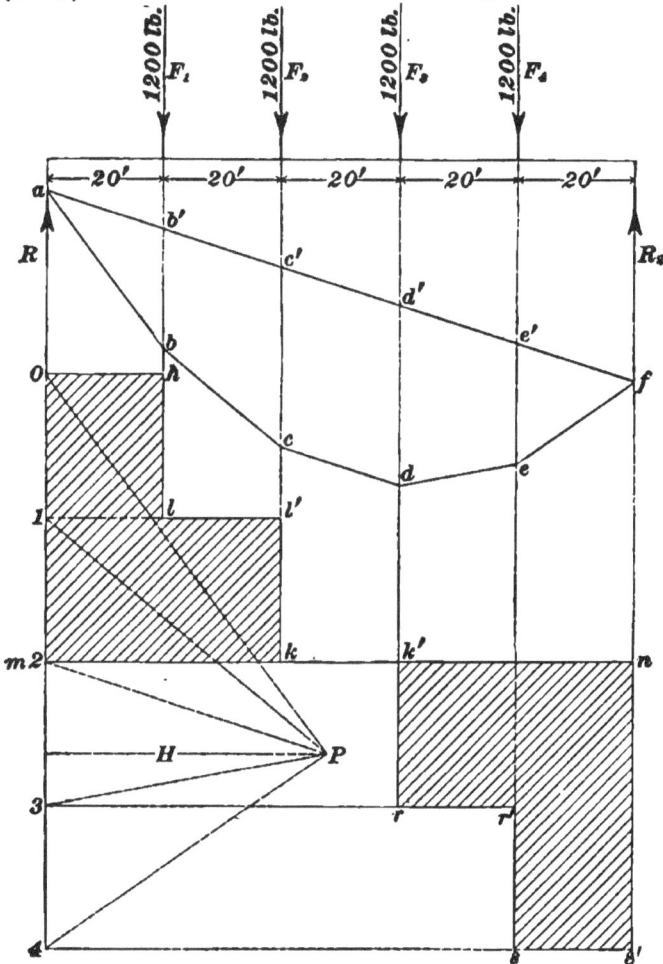

Scale of forces 1"=1600 lb.
Scale of distance 1"=32'
FIG. 54.

moments and shear diagram is clearly shown in Fig. 54.   It
is so nearly like that of Fig. 46 that a detailed description
is unnecessary.   It will be noticed that between k and k'
the shear is zero, and that since the reactions are equal the
shear at either support $= \frac{1}{2}$ of the load $= 2,400$ lb.   The
greatest intercept is $c\,c' = d\,d' = 30$ ft.   The pole distance
$H = 2,400$ lb.   Hence, the bending moment $= 2,400 \times 30 =$
72,000 ft.-lb. $= 72,000 \times 12 = 864,000$ in.-lb.

SURVEYING.

---

(614) Let x = number of degrees in angle C; then, $2x$ = angle A and $3x$ = angle B. The sum of A, B, and C is $x + 2x + 3x = 6x = 180°$, or $x = 30° = C$; $2x = 60° = A$, and $3x = 90° = B$. Ans.

(615) Let x = number of degrees in one of the equal angles; then, $2x$ = their sum, and $2x \times 2 = 4x$ = the greater angle. $2x + 4x = 6x$ = sum of the three angles = $180°$; hence, $x = 30°$, and the greater angle = $30° \times 4 = 120°$. Ans.

(616) AB in Fig. 55 is the given diagonal 3.5 in. $3.5^2 = 12.25 \div 2 = 6.125$ in. $\sqrt{6.125} = 2.475$ in. = side of the required square. From A and B as centers with radii equal to 2.475 in., describe arcs intersecting at C and D. Connect the extremities A and B with the points C and D by straight lines. The figure $ACBD$ is the required square.

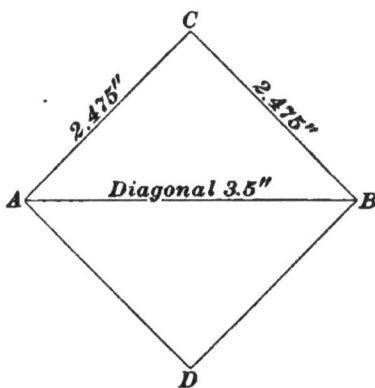

FIG. 55.

(617) Let AB, Fig. 56, be the given shorter side of the rectangle, 1.5 in. in length. At A erect an indefinite perpendicular AC to the line AB. Then, from B as a center with a radius of 3 in. describe an arc intersecting the perpendicular AC in the point D. This will give us two adjacent sides of the required rectangle. At B erect an indefinite perpendicular BE to AB, and at D erect an indefinite perpendicular DF to AD. These perpendiculars will intersect at G, and the resulting figure $ABGD$ will be the required rectangle. Its area is the product of the length AD by the width AB. $\overline{AD}^2 = \overline{BD}^2 - \overline{AB}^2$. $\overline{BD}^2 =$ 9 in.; $\overline{AB}^2 = 2.25$ in.; hence, $\overline{AD}^2 = 9$ in. $- 2.25$ in. $= 6.75$ in. $\sqrt{6.75} = 2.598$ in. $=$ side AD. 2.598 in. $\times 1.5 = 3.897$ sq. in., the area of the required rectangle.

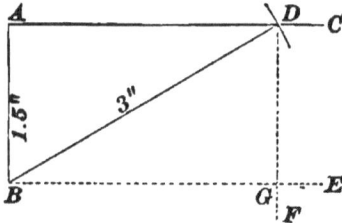

FIG. 56.

(**618**) (See Fig. 57.) With the two given points as centers, and a radius equal to $3.5'' \div 2 = 1.75' = 1\frac{3}{4}''$, describe short arcs intersecting each other. With the same radius and with the point of intersection as a center, describe a circle; it will pass through the two given points.

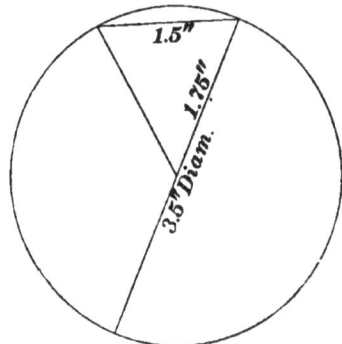

FIG. 57.

(**619**) AB in Fig. 58 is the given line, A the given point, AC and $CB = AB$. The angles A, B, and C are each equal to 60°. From B and C as centers with equal radii, describe arcs intersecting at D. The line AD bisects the angle A; hence, angle $BAD = 30°$.

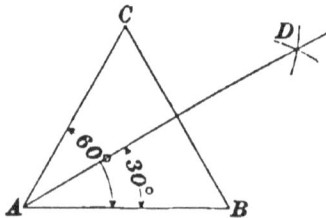

FIG. 58.

(**620**) Let AB in Fig. 59 be one of the given lines, whose length is 2 in., and let AC, the other line, meet AB at A, forming an angle of 30°. From A and B as centers, with radii equal to $A\,B$, describe arcs intersecting at D. Join $A\,D$ and $B\,D$. The triangle $A\,B\,D$ is equilateral; hence, each of its angles, as A, contains 60°. From D as a

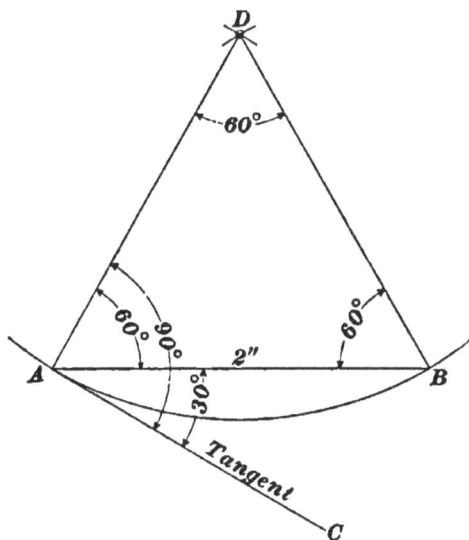

FIG. 59.

center, with a radius $A\,D$, describe the arc $A\,B$. The line $A\,C$ is tangent to this arc at the point A.

(621) $A\,B$ in Fig. 60 is the given line, C the given point, $C\,D = C\,E$. From D and E as centers with the same radii, describe arcs intersecting at F. Through F draw $C\,G$. Lay off $C\,G$ and $C\,B$ equal to each other, and from B and G as centers with equal radii describe arcs intersecting at H. Draw $C\,H$. The angle $B\,C\,H = 45°$.

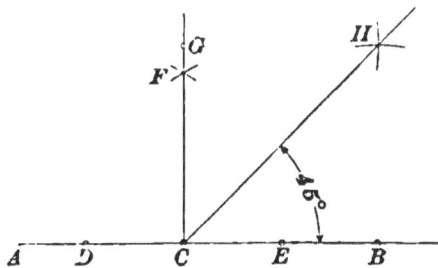

FIG. 60.

(622) 1st. A B in Fig. 61 is the given line 3 in. long. At A and B erect perpendiculars A C and B D each 3 in.

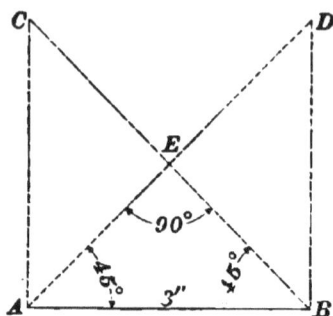

FIG. 61.

long. Join A D and B C. These lines will intersect at some point E. The angles $E A B$ and $E B A$ are each 45°, and the sides $A E$ and $E B$ must be equal, and the angle $A E B = 90°$.

2d. A B in Fig. 62 is the given line and c its middle point. On A B describe the semicircle A E B. At c erect a perpendicular to A B, cutting the arc $A E B$ in E. Join $A E$ and $E B$. The angle $A E B$ is 90°.

(623) See Art. **1181** and Fig. 237.

(624) See Art. **1187.**

(625) (See Art. **1195** and Fig. 245.) Draw the line

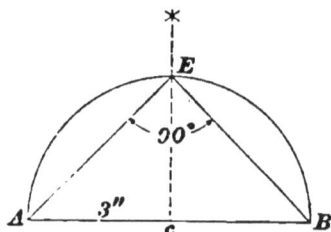

FIG. 62.

A B, Fig. 63, 5 in. long, the length of the given side. At A draw the indefinite line A C, making the angle B A C

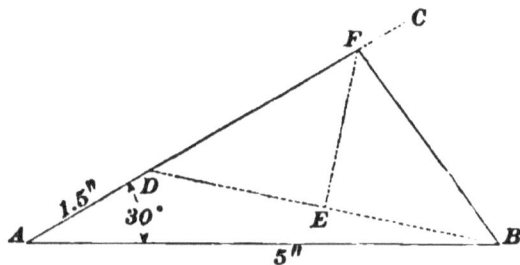

FIG. 63.

equal to the given angle of 30°. On A C lay off A D 1.5 in. long, the given difference between the other two sides of the triangle. Join the points *B* and *D* by a straight line, and at its middle point *E* erect a perpendicular, cutting the line *A C* in the point *F*. Join *B F*. The triangle *A B F* is the required triangle.

(626) See Art. **1198.**

(627) See Art. **1200.**

(628) See Art. **1201.**

(629) See Art. **1204.**

(630) See Art. **1205.**

(631) See Arts. **1205** and **1206.**

(632) See Art. **1206.**

(633) See Art. **1204.**

(634) See Art. **1207.**

(635) See Art. **1207.**

(636) See Arts. **1209** and **1213.**

(637) See Art. **1211.**

(638) (*a*) In this example the *declination* is *east*, and

Magnetic Bearing.	True Bearing.
N 15° 20′ E	N 18° 35′ E
N 88° 50′ E	S 87° 55′ E
N 20° 40′ W	N 17° 25′ W
N 50° 20′ E	N 53° 35′ E

for a course whose *magnetic bearing* is N E or S W, the

true bearing is the *sum* of the magnetic bearing and the declination. For a course whose *magnetic bearing* is N W or S E, the *true bearing* is the difference between the magnetic bearing and the declination.

As the first magnetic bearing is N 15° 20′ E, the true bearing is the *sum* of the *magnetic bearing* and the *declination*. We accordingly make the addition as follows:

$$\begin{array}{r} \text{N } 15° \ 20′ \text{ E} \\ 3° \ 15′ \\ \hline \text{N } 18° \ 35′ \text{ E,} \end{array}$$

and we have N 18° 35′ E as the true bearing of the first course.

The second magnetic bearing is N 88° 50′ E, and we add the declination of 3° 15′ to that bearing, giving N 92° 05′ E. This takes us past the east point to an amount equal to the *difference* between 90° and 92° 05′, which is 2° 05′. This angle we *subtract* from 90°, the total number of degrees between the south and east points, giving us S 87° 55′ E for the *true bearing* of our line. A simpler method of determining the true bearing, when the sum of the magnetic bearing and the declination exceeds 90°, is to subtract that sum from 180°; the difference is the true bearing. Applying this method to the above example, we have 180° — 92° 05′ = S 87° 55′ E.

The third magnetic bearing is N 20° 30′ W, and the *true bearing* is the *difference* between that bearing and the *declination*. We accordingly deduct from the magnetic bearing N 20° 40′ W, the declination 3° 15′, which gives N 17° 25′ W for the true bearing.

(*b*) Here the *declination* is *west*, and for a course whose *magnetic bearing* is N W or S E the *true bearing* is the *sum* of the magnetic bearing and the declination. For a course whose *magnetic bearing* is N E or S W, the *true bearing* is the *difference* between the magnetic bearing and the declination.

The first magnetic bearing is N 7° 20' W, and as the declination is west, it will be added. We, therefore, have for the true bearing N 7° 20' W + 5° 10' = N 12° 30' W.

Magnetic Bearing.	True Bearing.
N 7° 20' W	N 12° 30' W
N 45° 00' E	N 39° 50' E
S 15° 20' E	S 20° 30' E
S 2° 30' W	S 2° 40' E

The next two bearings the student can readily determine for himself.

The fourth magnetic bearing is S 2° 30' W, and to obtain the true bearing we must *subtract* the declination, i. e., we must change the direction eastwards. A change of 2° 30' will bring us *due south;* hence, the bearing will be east of south to an amount equal to the difference between 2° 30' and the total declination 5° 10', which is 2° 40'. The true bearing is, therefore, S 2° 40' E.

(639) See Art. **1216.**

(640) See Art. **1217.**

(641) See Art. **1219.**

(642) A plat of the accompanying notes is given in Fig. 64 to a scale of 600 ft. to the inch. The order of work is as follows:

First draw a meridian $N S$ (see Fig. 64), and then assume the starting point A, which call Sta. 0. Through A draw a meridian $A B$ parallel to $N S$. Then, placing the center of the protractor at A, with its zero point in the line $A B$, lay off the bearing angle 10° 10' to the right of

A B, as the bearing is N E. Mark the point of angle measurement carefully, and draw a line joining it and the point *A*. This line will give the direction of the first course, the end of which is at Sta. 5 + 20, giving 520 ft. for the length of that course. On this line lay off to a scale of 600 ft. to the inch the distance 520 ft., locating the point *C*, which is Sta. 5 + 20. Through *C* draw a meridian *C D*, and with the protractor lay off the bearing angle N 40° 50' E to the right of the meridian, marking the point of angle measurement and joining it with the point *C* by a straight line, which will be the direction of the second course. The end of this course is Sta. 10 + 89, and its length is the difference between 1,089 and 520, which is

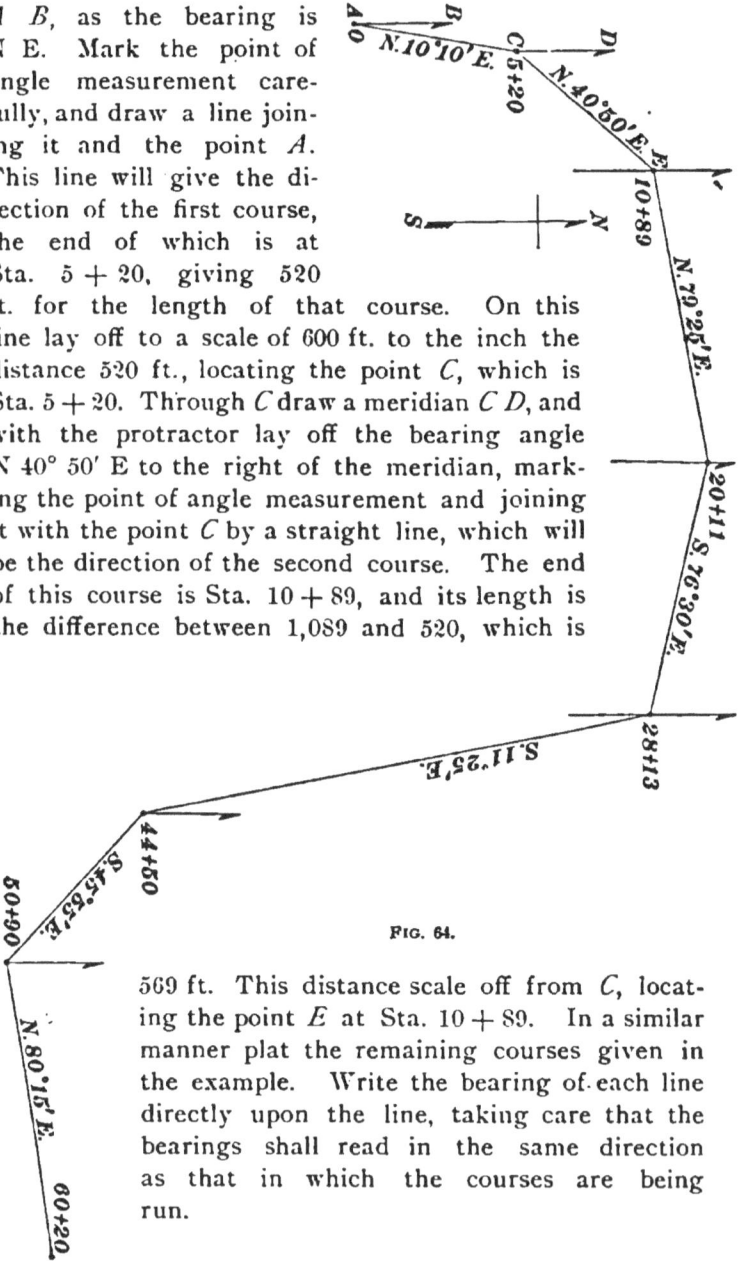

FIG. 64.

569 ft. This distance scale off from *C*, locating the point *E* at Sta. 10 + 89. In a similar manner plat the remaining courses given in the example. Write the bearing of each line directly upon the line, taking care that the bearings shall read in the same direction as that in which the courses are being run.

(643) See Art. **1231** and Figs. 261 and 262.

(644) For *first* adjustment see Art. **1233.** For *second* adjustment see Art. **1234** and Fig. 263, and for *third* adjustment see Art. **1235** and Fig. 264.

(645) See Art. **1238** and Fig. 265.

(646) See Art. **1239** and Fig. 266.

(647) See Art. **1240** and Fig. 267.

(648) See Art. **1242** and Fig. 269.

(649) See Art. **1242** and Fig. 269.

(650) To the bearing at the given line, viz., N 55° 15′ E, we add the angle 15° 17′, which is turned to the right. This gives for the second line a bearing of N 70° 32′ E.

(651) To the bearing of the given line, viz., N 80° 11′, we add the angle 22° 13′, which is the amount of change in the direction of the line. The sum is 102° 24′, and the direction is 102° 24′ to the *right* or *east* of the *north* point of the compass. At 90° to the right of north the direction is due east. Consequently, the direction of the second line must be *south* of east to an amount equal to the difference between 102° 24′ and 90°, which is 12° 24′. Subtracting this angle from 90°, the angle between the south and east points, we have 77° 36′, and the direction of the second line is S 77° 36′ E. The simplest method of determining the direction of the second line is to subtract 102° 24′ from 180° 00′. The difference is 77° 36′, and the direction changing from N E to S E gives for the second line a bearing of S 77° 36′ E.

(652) To the bearing of the given line, viz., N 13° 15′ W, we add the angle 40° 20′, which is turned to the left. The sum is 53° 35′, which gives for the second line a bearing of N 53° 35′ W.

(653) The bearing of the first course, viz., S 10° 15′ W, is found in the column headed Mag. Bearing, opposite Sta. 0. For the first course, the deduced or calculated bearing must be the same as the magnetic bearing. At Sta. 4 + 40,

an angle of 15° 10′ is turned to the right. It is at once evident that if a person is traveling in the direction S 10° 15′ W,
and changes his course to the right 15° 10′, his course will
approach a due westerly direction by the amount of the
change, and the direction of his second course is found by
adding to the first course, viz., S 10° 15′, the amount of such

Station.	Deflection.	Mag. Bearing.	Ded. Bearing.
54 + 25			
49 + 20	L. 25° 14′	S 25° 40′ W	S 25° 39′ W
44 + 80	L. 10° 47′	S 50° 50′ W	S 50° 53′ W
33 + 77	R. 16° 55′	S 61° 45′ W	S 61° 40′ W
25 + 60	R. 24° 40′	S 44° 50′ W	S 44° 45′ W
16 + 20	L. 15° 35′	S 20° 00′ W	S 20° 05′ W
8 + 90	R. 10° 15′	S 35° 50′ W	S 35° 40′ W
4 + 40	R. 15° 10′	S 25° 20′ W	S 25° 25′ W
0		S 10° 15′ W	S 10° 15′ W

change in direction. The sum is 25° 25′, and the second
course S 25° 25′ W. The needle at this point reads S 25° 20′ W.
The difference between the magnetic bearing and the calculated bearing may be owing to local attraction, but as we
can not read the needle to within 10 minutes, we must gen-
erally ascribe small discrepancies to that cause. This calculated bearing we write in its proper column opposite Sta.
4 + 40, where the change in direction occurred.

The next angle is 10° 15′ to the right, which we add to the
previous calculated bearing S 25° 25′ W, giving S 35° 40′ W
for the calculated bearing of the third course, which extends
from Sta. 8 + 90 to 16 + 20. In a similar manner, the student will calculate the remaining bearings, considering well
how the changes in direction will affect his relations to the
points of the compass. A plat of the notes to a scale of
400 ft. to the inch is given in Fig. 65.

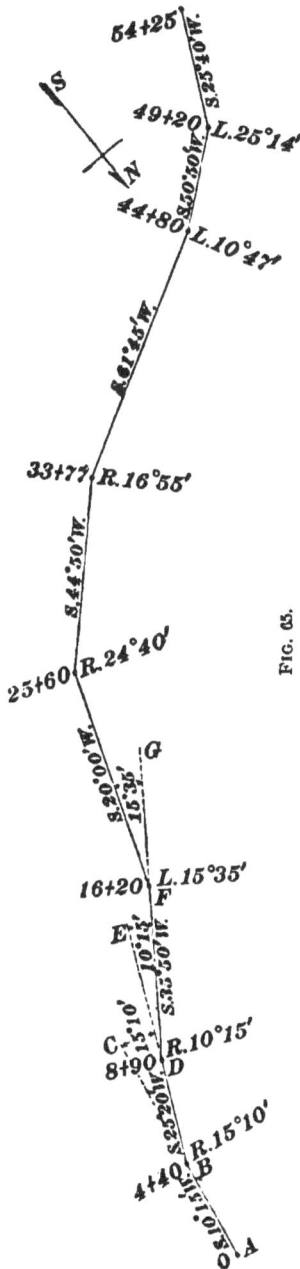

FIG. 65.

Assume the starting point A, Fig. 65, which is Sta. 0, and through it draw a straight line. The first angle, viz., 15° 10' to the right, is turned at Sta. 4 + 40, giving for the first course a length of 440 ft. Scale off this distance from $0 A$ to a scale of 400 ft. to the inch, locating the point B, which is Sta. 4 + 40, and write on the line its bearing S 10° 15' W, as recorded in the notes. Produce $A B$ to C, making $B C$ greater than the diameter of the protractor, and from B lay off to the right of $B C$ the angle 15° 10'. Join this point of angle measurement with B by a straight line, giving the direction of the second course. The end of this course is at Sta. 8 + 90, and its length is the difference between 8 + 90 and 4 + 40, which is 450 ft. This distance we scale off from B, Sta. 4 + 40, locating the point D, and write on the line its bearing of S 25° 20' W, as found in the notes. We next produce $B D$ to E, making $D E$ greater than the diameter of the protractor, and at D lay off the angle 10° 15' to the right of $D E$, giving the direction of the next course. In a similar manner, plat the remaining notes given in Example 653. The student in his drawing will show the prolongation of only the first three lines, drawing such prolongations in dotted lines. In platting the remaining angles, he will produce the lines in pencil only, erasing them as soon as the forward angle is laid off. Write the proper station number in pencil at the end of each line as soon as platted, and the angle with its direction, R. or L., before laying off the following angle. Write the bearing of each line distinctly, the letters reading in the same direction in which the line is being run. The magnetic meridian is platted as follows: The bearing of the course from Sta. $33 + 77$ to Sta. $44 + 80$ is S $61°$ $45'$ W, i. e., the course is $61°$ $45'$ to the left of a north and south line, which is the direction we wish to indicate on the map. Accordingly, we place a protractor with its center at Sta. $33 + 77$ and its zero on the following course, and read off the angle $61°$ $45'$ to the right. Through this point of angle measurement and Sta. $33 + 77$ draw a straight line N S. This line is the required meridian.

(654) Angle $B = 39°$ $25'$. From the principles of trigonometry (see Art. 1243), we have the following proportion:

sin $39°$ $25'$: sin $60'$ $15'$ :: 415 ft. : side A B.

sin $60°$ $15' = .8682$.

415 ft. $\times$ .8682 $= 360.303$ ft.

sin $39°$ $25' = .63496$.

$360.303 \div .63496 = 567.442$ ft., the side A B. Ans.

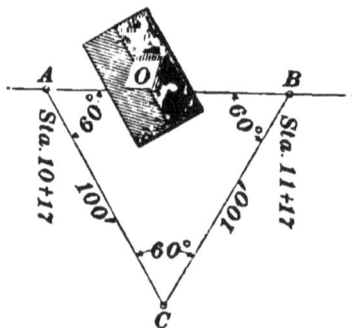

FIG. 66.

(655) (See Fig. 66.) At A we turn an angle B A C of $60°$ and set a plug at C $100'$ from A. At C we turn an angle A C B of $60°$ and set a plug at B $100'$ from C. The point B will be in the line A B, and setting up the instrument at B, we turn the angle C B $A = 60°$. The instrument is then reversed, and the line A B produced as required.

(656) See Art. **1245.**

(657) See Art. **1246.**

(658) See Art. **1246.**

(659) See Art. **1248.**

(660) See Art. **1249.**

(661) A $5°$ curve is one in which a central angle of $5°$ will subtend a chord of 100 ft. at its circumference. Its radius is practically one-fifth of the radius of a $1°$ curve, and equal to 5,730 ft. ÷ 5 = 1,146 ft.

(662) The degree of curve is always twice as great as the deflection angle.

(663) See Art. **1249** and Fig. 282.

(664) Formula **90,** $C = 2 R \sin D.$ (See Art. **1250.**)

(665) Formula **91,** $T = R \tan \frac{1}{2} I.$ (See Art. **1251.**)

(666) The intersection angle $C E F$, being external to the triangle $A E C$, is equal to the sum of the opposite interior angles A and C. $A = 22° 10'$ and $C = 23° 15'$. Their sum is $45° 25' = C E F.$ Ans.

The angle $A E C = 180° - (22° 10' + 23° 15') = 134° 35'$.

From the principles of trigonometry (see Art. **1243**), we have

$\sin 134° 35' : \sin 23° 15' :: 253.4$ ft. : side $A E$;

whence, side $A E = 140.44$ ft., nearly. Ans.

Also, $\sin 134° 35' : \sin 22° 10' :: 253.4$ ft. : side $C E$;

whence, side $C E = 134.24$ ft., nearly. Ans.

(667) We find the tangent distance T by applying formula **91,** $T = R \tan \frac{1}{2} I.$ (See Art. **1251.**) From the table of Radii and Deflections we find the radius of a $6° 15'$ curve = 917.19 ft.; $\frac{1}{2} I = \dfrac{35° 10'}{2} = 17° 35'$; $\tan 17° 35' = .3169$. Substituting these values in the formula, we have $T = 917.19 \times .3169 = 290.66$ ft. Ans.

(668) We find the tangent distance T by applying formula **91,** $T = R \tan \frac{1}{2} I.$ (See Art. **1251.**) From the table of Radii and Deflections we find the radius of a 3° 15′ curve is 1,763.18 ft.; $\frac{1}{2} I = \dfrac{14°\ 12′}{2} = 7°\ 06′$; tan $7°\ 06′ =$ .12456. Substituting these values in the above formula, we have $T = 1,763 \times .12456 = 219.62$ ft.   Ans.

(669) See Art. **1252.**

(670) The angle of intersection 30° 45′, reduced to decimal form, is 30.75°. The degree of curve 5° 15′, reduced to decimal form, is 5.25°. Dividing the intersection angle 30.75° by the degree of curve 5.25 (see Art. **1252**), the quotient is the required length of the curve in stations of 100 ft. each. $\dfrac{30.75°}{5.25°} = 5.8571$ full stations equal to 585.71 ft.

(671) In order to determine the P. C. of the curve, we must know the tangent distance which, subtracted from the number of the station of the intersection point, will give us the P. C. We find the tangent distance T by applying formula **91**, $T = R$ tan $\frac{1}{2} I$. (See Art. **1251.**) From the table of Radii and Deflections we find the radius of a 5° curve is 1,146.28 ft.; $\frac{1}{2} I = \dfrac{33°\ 06′}{2} = 16°\ 33′$; tan 16° 33′ = .29716.

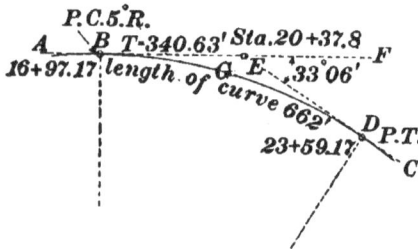

FIG. 67.

Substituting these values in formula **91**, we have $T = 1,146.28 \times .29716 = 340.63$ ft. In Fig. 67, let $A\ B$ and $C\ D$ be the tangents which intersect in the point E, forming an angle $D\ E\ F = 33°\ 06′$. The line of survey is being run in the direction $A\ B$, and the line is measured in regular order up to the intersection point E, the station of which is 20 + 37.8. Subtracting the tangent distance, $B\ E = 340.63$ ft. from Sta. 20 + 37.8, we have 16 + 97.17, the station of the P. C. at B. The intersection angle 33° 06′ in decimal form is 33.1°. Dividing this angle by 5, the degree of the curve, we obtain the length $B\,G\,D$ of the curve in full stations. $\frac{33.1}{5}=6.62$ stations $= 662$ ft. The length of the curve, 662 ft., added to the station of the P. C., viz., $16 + 97.17$, gives $23 + 59.17$, the station of the P. T. at D.

(672) The given tangent distance, viz., 291.16 ft., was obtained by applying formula **91**, $T = R \tan \frac{1}{2} I$ (see Art. **1251**), $I = 20°\ 10'$, and $\frac{1}{2} I = 10°\ 05'$, tan $10°\ 05' = .17783$. Substituting these values in the above formula, we have $291.16 = R \times .17783$; whence, $R = \frac{291.16}{.17783} = 1,637.29$ ft. Ans.

The degree of curve corresponding to the radius 1,637.29 we determined by substituting the radius in formula **89**, $R = \frac{50}{\sin D}$ (see Art. **1249**), and we have

$$1,637.29 = \frac{50}{\sin D}; \text{ whence, } \sin D = \frac{50}{1,637.29} = .03054.$$

The deflection angle corresponding to the sine .03054 is $1°\ 45'$, and is one-half the degree of the curve. The degree of curve is, therefore, $1°\ 45' \times 2 = 3°\ 30'$. Ans.

(673) Formula **92**, $d = \frac{c^2}{R}$. (See Art. **1255** and Fig. 283.)

(674) The ratio is 2; i. e., the chord deflection is double the tangent deflection. (See Art. **1254** and Fig. 283.)

(675) As the degree of the curve is $7°$, the deflection angle is $3°\ 30' = 210'$ for a chord of 100 ft., and for a chord of 1 ft. the deflection angle is $\frac{210'}{100} = 2.1'$; and for a chord of 48.2 ft. the deflection angle is $48.2 \times 2.1' = 101.22' = 1°\ 41.22'$.

(676) The deflection angle for 100-ft. chord is $\frac{6°\ 15'}{2} = 3°\ 07\frac{1}{2}' = 187.5'$, and the deflection angle for a 1-ft. chord is

$\dfrac{187.5'}{100} = 1.875'$. The deflection angle for a chord of 72.7 ft:

is, therefore, $1.875' \times 72.7 = 136.31' = 2° 16.31'$.

(**677**) We find the tangent deflection by applying formula **93**, tan def. $= \dfrac{c^2}{2\,R}$. (See Art. **1255**.) $c = 50$. $50^2 = 2,500$. The radius R of $5° 30'$ curve $= 1,042.14$ ft. (See table of Radii and Deflections.) Substituting these values in formula **93**, we have tan def. $= \dfrac{2,500}{2,084.28} = 1.199$ ft. Ans.

(**678**) The formula for chord deflections is $d = \dfrac{c^2}{R}$. (See Art. **1255**, formula **92**.) $c = 35.2$. $35.2^2 = 1,239.04$. The radius R of a $4° 15'$ curve is $1,348.45$ ft. Substituting these values in formula **92**, we have $d = \dfrac{1,239.04}{1,348.45} = .919$ ft. Ans.

(**679**) The formula for finding the radius R is $R = \dfrac{50}{\sin D}$. (See Art. **1249**.) The degree of curve is $3° 10'$. D, the deflection angle, is $\dfrac{3° 10'}{2} = 1° 35'$; $\sin 1° 35' = .02763$. Substituting the value of $\sin D$ in the formula, we have $R = \dfrac{50}{.02763}$; whence, $R = 1,809.63$ ft. Ans.

The answer given with the question, viz., 1,809.57 ft., agrees with the radius given in the table of Radii and Deflections, which was probably calculated with sine given to eight places instead of five places, as in the above calculation, which accounts for the discrepancy in results.

(**680**) In Fig. 68, let $A\,B$ and $A\,C$ represent the given lines, and $B\,C$ the amount of their divergence, viz., 18.22

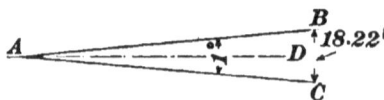

ft. The lines will form a triangle $A\,B\,C$, of which the angle $A = 1°$. Draw a perpendicular from A to

FIG. 68.

D, the middle point of the base. The perpendicular will

bisect the angle A and form two right angles at the base of the triangle. In the triangle ADB we have, from rule **5**, Art. **754**, $\tan BAD = \dfrac{BD}{AD}$. $BAD = 30'$, $BD = \dfrac{18.22 \text{ ft.}}{2} =$ 9.11 ft., and $\tan 30' = .00873$. Substituting known values in the equation, we have $.00873 = \dfrac{9.11}{AD}$; whence, $AD = \dfrac{9.11 \text{ ft.}}{.00873} = 1,043.53$ ft. Ans.

By a practical method, we determine the length of the lines by the following proportion:

$$.1.745 : 18.22 :: 100 \text{ ft.} : \text{the required length of line;}$$

whence, length of line $= \dfrac{1,822}{1.745} = 1,044.13$ ft. Ans.

The second result is an application of the principle of two lines 100 ft. in length forming an angle of 1° with each other, which will at their extremity diverge 1.745 ft.

(**681**) Degree of curve $= \dfrac{24° \ 15'}{6.0625} = \dfrac{24.25°}{6.0625} = 4°$. Ans.

(**682**) See Arts. **1264, 1265, 1266,** and **1267.**

(**683**) See Arts. **1269 -1274.**

(**684**) Denote the radius of the bubble tube by x; the distance of the rod from the instrument, viz., 300', by d; the difference of rod readings, .03 ft., by h, and the movement of the bubble, viz., .01 ft., by S. By reference to Art. **1275** and Fig. 289, we will find that the above values have the proportion $h : S :: d : x$. Substituting known values in the proportion, we have $.03 : .01 :: 300 : x$; whence, $x = \dfrac{3}{.03} = 100$ ft., the required radius. Ans.

(**685**) See Art. **1277.**

(**686**) See Art. **1278.**

(**687**) To the elevation 61.84 ft. of the given point, we add 11.81 ft., the backsight. Their sum, 73.65 ft., is the height of instrument. From this H. I., we subtract the foresight to the T. P., viz., 0.49 ft., leaving a difference of 73.16 ft., which is the elevation of the T. P. (See Art. **1279.**)

(**688**)   See Art. **1280.**

(**689**)   See Art. **1281.**

. (**690**)   See Art. **1282.**

(**691**)   See Art. **1286.**

(**692**)   See Art. **1289.**

(**693**)   See Art. **1290.**

(**694**)   See Art. **1291.**

(**695**)   The distance between Sta. 66 and Sta. 93 is 27 stations.   As the rate of grade is $+ 1.25$ ft. per station, the total rise in the given distance is 1.25 ft. $\times 27 = 33.75$ ft., which we add to 126.5 ft., the grade at Sta. 66, giving 160.25 ft. for the grade at Sta. 93.   (See Art. **1291.**)

(**696**)   See Art. **1292.**

(**697**) $\dfrac{-16.4'}{56'}$ $\dfrac{-10.3'}{73'}$ $\Big|$ $\dfrac{+11.4'}{84'}$ $\dfrac{+8.8'}{96'}$

Contour 50.0 at 48.5 ft. to left of Center Line. | Contour 60.0 at 26 ft. to right of Center Line.
Contour 40.0 at 94.0 ft. to left of Center Line. | Contour 70.0 at 106.9 ft. to right of Center Line.
Contour 30.0 at 128.0 ft. to left of Center Line. | Elevation 76.7 at 180 ft. to right of Center Line.

(**698**)   The elevations of the accompanying level notes are worked out as follows: The first elevation recorded in the column of elevations is that of the *bench mark*, abbreviated to B. M.   This elevation is 161.42 ft.   The first rod reading, 5.53 ft., is the backsight on this B. M., a *plus* reading, and recorded in column of rod readings.   This rod reading we add to the elevation of the bench mark, to determine the height of instrument, as follows: 161.42 ft. + 5.53 ft. = 166.95 ft., the H. I.   The next rod reading, which is at Sta. 40, is 6.4 ft.   The rod reading means that the surface of the ground at Sta. 40 is 6.4 ft. below the horizontal axis of the telescope.   The elevation of that surface is, therefore, the difference between 166.95 ft., the H. I., and 6.4 ft., the rod reading.   166.95 − 6.4 = 160.55 ft.   The $\dfrac{5}{100}$ ft. is a fraction so small that in surface elevations it is the universal practice to ignore it, and the elevation of the ground at

Station.	Rod Reading.	Height Instrument.	Elevation.	Grade.
B. M.	+ 5.53	166.95	161.42	
40	6.4		160.5	162.0
41	7.2		159.7	160.485
41 + 60	10.9		.156.0	
42	8.6		158.3	158.97
43	8.8		158.1	157.455
T. P. −	8.66		158.29	
+	2.22	160.51		
44	4.8		155.7	155.94
45	6.3		154.2	154.425
46	8.8		151.7	152.91
47	9.9		150.6	151.395
48	11.1		149.4	149.88
T. P. −	11.24		149.27	
+	3.30	152.57		
49	4.7		147.9	148.365
50	7.1		145.5	146.85
51	8.7		143.9	145.335
52	9.8		142.8	143.82
53	10.9		141.7	142.305
T. P. −	11.62		140.95	

Sta. 40 is taken at 160.5 ft. The rod reading at Sta. 41 is 7.2, which, subtracted from 166.95 ft., gives for that station an elevation of 159.7. The remaining rod readings up to

and including that at Sta. 43, we subtract from the same
H. I., viz., 166.95. Here at a turning point (T. P.) of 8.66
ft. is taken and recorded in the column of rod readings.
This reading being a foresight is *minus*, and is subtracted
from the preceding H. I. This gives us for the elevation of
the T. P., 166.95 ft. − 8.66 ft. = 158.29 ft., which we record in
the column of elevations. The instrument is then moved forwards and a backsight of 2.22 ft taken on the same T. P. and
recorded in the column. This is a *plus* reading, and is added
to the elevation of the T. P., giving us for the next H. I.
an elevation of 158.29 ft. + 2.22 ft. = 160.51 ft. The next
rod reading, viz., 4.8, is at Sta. 44, and the elevation at that
station is the difference between the preceding H. I., 160.51,
and that rod reading, giving an elevation of 160.51 ft. − 4.8
ft. = 155.7 ft., which is recorded in the column of elevations
opposite Sta. 44. In a similar manner, the remaining elevations are determined.

In checking level notes, only the *turning points* rod readings are considered. It will be evident that starting from
a given bench mark, all the *backsight* or *plus* readings will
add to that elevation, and all the *foresight* or *minus* readings will subtract from that elevation. If now we place in
one column the height of the B. M., together with all the ·
backsight or + readings, and in another column all the foresight or − readings, and find the sum of each column, then,
by subtracting the sum of the − readings from the sum of
the + readings, we shall find the elevation of the last point
calculated, whether it be a turning point or a height of instrument. Applying this method to the foregoing notes
we have the following:

	+ readings.	− readings.
B. M.	161.42 ft.	8.66 ft.
	5.53 ft.	11.24 ft.
	2.22 ft.	11.62 ft.
	3.30 ft.	31.52 ft.
	172.47 ft.	
	31.52 ft.	
	140.95 ft.	

The difference of the columns, viz., 140.95 ft., agrees with the elevation of the T. P. following Sta. 53, which is the last one determined. A check mark ✓ is placed opposite the elevation checked, to show that the figures have been verified. The rate of grade is determined as follows: In one mile there are 5,280 ft. = 52.8 stations. A descending grade of 80 ft. per mile gives per station a descent of $\dfrac{80 \text{ ft.}}{52.8} = 1.515$ ft. The elevation of the grade at Sta. 40 is fixed at 162.0 ft. As the grade descends from Sta. 40 at the rate of 1.515 ft. per station, the grade at Sta. 41 is found by subtracting 1.515 ft. from 162.0 ft., which gives 160.485 ft., and the grade for each succeeding station is found by subtracting the rate of grade from the grade of the immediately preceding station.

A section of profile paper is given in Fig. 69 in which the level notes are platted, and upon which the given grade line

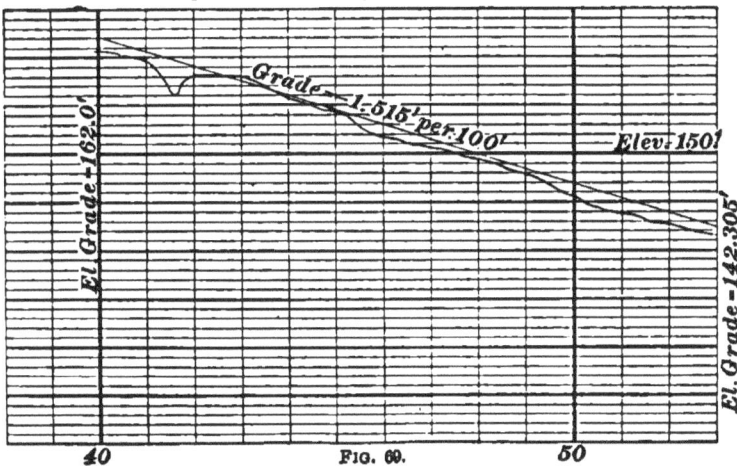

FIG. 69.

is drawn. The profile is made to the following scales: viz., horizontal, 400 ft. = 1 in.; vertical, 20 ft. = 1 in.

Every fifth horizontal line is heavier than the rest, and each twenty-fifth horizontal line is of *double weight*. Every tenth vertical line is of *double weight*. The spaces between the vertical lines represent 100 ft., and those between the horizontal lines 1 ft. The figure represents 1,500 ft. in length and 45 ft. in height. Assume the elevation of the sixth heavy line from the bottom at 150 ft. The second vertical line from the left of the figure is Sta. 40, which is written in the margin at the bottom of the page. Under the next heavy vertical line, ten spaces to the right, Sta. 50 is written. The elevation of Sta. 40, as recorded in the notes, is 160.5 ft. We determine the corresponding elevation in the profile as follows:

As the elevation of the sixth heavy line from the bottom is assumed at 150 ft., 160.5 ft., which is 10.5 ft. higher, must be $10\frac{1}{2}$ spaces above this line. This additional space covers two heavy lines and one-half the next space. This point is marked in pencil. The elevation of Sta. 41, viz., 159.7 ft., we locate on the next vertical line and 9.7 spaces above the 150 ft. line. The next elevation, 156.0 ft., is at Sta. 41 + 60. This distance of 60 ft. from Sta. 41 we estimate by the eye and plat the elevation in its proper place. In a similar manner, we plat the remaining elevations and connect the points of elevation by a continuous line drawn free-hand. The grade at Sta. 40 is 162.0 ft. This elevation should be marked in the profile by a point enclosed by a small circle. At each station between Sta. 40 and Sta. 53 there has been a descent of 1.515 ft., making a total descent between these stations of $1.515 \times 13 = 19.695$. The grade at Sta. 53 will, therefore, be 162.0 ft. $- 19.695$ ft. $= 142.305$ ft. Plat the elevation in the profile at Sta. 53, and enclose the point in a small circle. Join the grade point at Sta. 40 with that at Sta. 53 by a straight line, which will be the grade·line required. Upon this line mark the grade $- 1.515$ per 100 ft.

(699) $\quad \dfrac{-5°}{65'} \quad \dfrac{-9°}{117'} \quad \Big| \quad \dfrac{+11°}{120'}$

Nine 5-foot contours are included within the given slopes, as follows:

Contour 70.0 at 31.5 ft. to left of Center Line.	Contour 80.0 at 25.5 ft. to right of Center Line.
Contour 65.0 at 63.0 ft. to left of Center Line.	Contour 85.0 at 51.0 ft. to right of Center Line.
Contour 60.0 at 94.5 ft. to left of Center Line.	Contour 90.0 at 76.5 ft. to right of Center Line.
Contour 55.0 at 133.0 ft. to left of Center Line.	Contour 95.0 at 102.0 ft. to right of Center Line.
Elevation 50.7 at 182.0 ft. to left of Center Line.	Elevation 98.5 at 120.0 ft. to right of Center Line.

(700) 1.745 ft. × 3 = 5.235 ft., the vertical rise of a 3° slope in 100 ft., or 1 station and $\dfrac{10}{5.235} = 1.91$ stations $= 191$ ft.

(701) (See Question 701, Fig. 15.) From the instrument to the center of the spire is 100 ft. + 15 = 115 ft., and we have a right triangle whose base $A\,D = 115$ ft. and angle A is 45° 20′. From rule **5**, Art. **754**, we have tan 45° 20′ $=$ $\dfrac{\text{side }B\,D}{115}$; whence, $1.01170 = \dfrac{\text{side }B\,D}{115}$; or $B\,D = 116.345$ feet. The instrument is 5 feet above the level of the base; hence, 116.345 ft. + 5 ft. = 121.345, the height of the spire.

(702) Apply formula **96**.

$$Z = (\log h - \log H) \times 60{,}384.3 \times \left(1 + \frac{t + t' - 64°}{900}\right),$$

(See Art. **1304**.)

$$\log \text{ of } h,\ 29.40 = 1.46835$$
$$\log \text{ of } H,\ 26.95 = 1.43056$$
$$\overline{\text{Difference} = 0.03779}$$

$$1 + \frac{t + t' - 64°}{900} = 1 + \frac{74 + 58 - 64}{900} = 1.0755.$$

Hence, $Z = .03779 \times 60{,}384.3 \times 1.0755 = 2{,}454$ ft., the difference in elevation between the stations.

(703) See Art. **1305**.

(704) See Art. **1308**.

(705) See Art. **1308**.

LAND SURVEYING

---

(706) See Art. **1309.**

(707) See Art. **1309.**

(708) See Art. **1310.**

(709) See Art. **1312.**

(710) See Art. **1313.**

(711) See Art. **1310** and Fig. 306.

(712) See Art. **1314** and Fig. 308.

(713) See Art. **1314** and Fig. 309.

(714) See Art. **1314.**

(715) See Art. **1315** and Fig. 310.

(716) See Art. **1317.**

(717) See Art. **1318.**

(718) See Art. **1319.**

(719) See Art. **1319** and Figs. 311 and 312.

(720) See Art. **1319.**

(721) See Art. **1319.**

(**722**) See Art. **1319.**

(**723**) See Art. **1320.**

(**724**) See Art. **1320.**

(**725**) See Art. **1321.**

(**726**) The magnetic variation is determined by subtracting the present bearing from B to C, viz., N 60° 15' E from the original bearing, viz., N 62° 00' E. The difference is 1° 45' and a *west* variation; hence, to determine the present bearings of the boundaries we must *add* the *variation* to an original bearing, which was N W or S E, and *subtract* it from an original bearing, which was N E or S W. The corrected bearings will be as follows:

Stations.	Original Bearings.	Distances.	Corrected Bearings.
A	N 31½° W	10.4 chains	N 33¼° W
B	N 62° E	9.2 chains	N 60° 15' E
C	S 36° E	7.6 chains	S 37¾° E
D	S 45½° W	10.0 chains	S 43¾° W

(**727**) See Art. **1323.**

(**728**) See Art. **1324.**

(**729**) See Art. **1326** and Fig. 316.

(**730**) See Art. **1327** and Figs. 317 and 318.

(**731**) See Art. **1328** and Fig. 319.

(**732**) As the bearing of the line $A B$ is N E, the end B will be east of the meridian passing through A. The departure of AB is the distance which B is east of A, or of the meridian passing through A. Now, if from B we drop a perpendicular BC upon that meridian, BC will be the departure of AB. The latitude of AB is the distance which the end B is north of the end A. The distance AC, measured on the meridian from A to the foot of the perpendicular from B, is the latitude of AB.

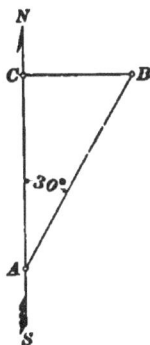

From an inspection of Fig. 70, we see that the line AB, together with its latitude AC and departure BC, form a right triangle, right angled at C, of which triangle BC is the sin and AC the cos of the bearing 30°. From rule 3, Art. **754**, we have

FIG. 70.

$\cos A = \dfrac{AC}{AB}$; whence, $AC = AB \cos A$; and from

rule 1, Art. **754**, $\sin A = \dfrac{BC}{AB}$; whence, $BC = AB \sin$

A, and we deduce the following:

Latitude = distance × cos bearing.
Departure = distance × sin bearing.

(733) See Art. **1329.**

(734)

Bearing.	Distances.	Latitudes.	Departures.
23¼°	400 ft.	3 6 7 5	1 5 7 9
	20 ft.	1 8 3 8	0 7 8 9
	3 ft.	2 7 5 6	1 1 8 4
	423 ft.	3 8 8.6 3 6 ft.	1 6 6.9 7 4 ft.

We divide the distance 423 ft. into three parts, viz., 400 ft., 20 ft., and 3 ft. If now we find the latitude and departure for 4 ft. and multiply them by 100, we shall obtain the latitude and departure for 400 ft. The latitude of 4 ft. is 3.675 ft., and the departure 1.579 ft. We place these figures under their proper headings as whole numbers. The latitude and departure of 2 ft. are 1.838 and 0.789, respectively, which we place as whole numbers under their proper headings, but removed one place to the right of the figures above them, as they are the latitude and departure of tens of feet. The latitude and departure of 3 ft. are 2.756 ft. and 1.184 ft., respectively, and we place them under their proper headings, but removed one place to the right as they are for units of feet. We now add up the partial latitudes and departures, and from the right of each sum we point off three decimal places, the same number as given in the traverse table, giving us for for the required latitude 388.636 ft., and for the required departure 166.974 ft.

(735)

Bearing.	Distances.	Latitudes.	Departures.
40°	200 ft.	1 5 3 2	1 2 8 6
	20 ft.	1 5 3 2	1 2 8 6
	5 ft.	3 8 3 0	3 2 1 4
	225 ft.	1 7 2.3 5 0 ft.	1 4 4.6 7 4 ft.

For the given bearing of 40° and distance of 225 ft., the latitude is 172.35 ft., and the *departure* 144.674 ft.

The complement of the given bearing is the difference between 90° and 40°, which is 50°. With this complement as the bearing, we have

Bearing.	Distances.	Latitudes.	Departures.
50°	200 ft.	1 2 8 6	1 5 3 2
	20 ft.	1 2 8 6	1 5 3 2
	5 ft.	3 2 1 4	3 8 3 0
	225 ft.	1 4 4.6 7 4 ft.	1 7 2.3 5 0 ft.,

in which the latitude and departure are exactly the reverse of those when the line had a bearing of 40°, the comple-ment of 50°.

(736) See Art. **1330.**

(737) We rule 11 columns, headed as below. The lati-tudes and departures for the several courses we calculate by traverse tables; placing the *north* latitudes, which are +, in the column headed N +, and the *south* latitudes, which are —, in the column headed S — ; the *east* departures, which are

+, in the column headed E +, and the *west* departures, which are —, in the column headed W —. These several columns we add, placing their sums at the foot of the columns. The sum of the distances is 37.20 chains; the sum of the north latitudes 13.19 chains, and of the south latitudes 13.16 chains. The difference is .03 chain, or 3 links. The sum of the east departures is 12.60 chains, and the sum of the west departures is 12.56 chains. The difference is .04 chain, or 4 links.

This difference indicates an error in either the bearings or measurements of the line or both. For had the work been correct, the sums of the north and south latitudes would have been equal. (See Art. **1330.**) The corrections for latitudes and departures are made as shown in the following proportions, the object of such correction being to make the sums of the north and south latitudes and of the east and west longitudes equal, and is called balancing the survey. (See Art. **1331.**)

Sta-tions.	Bearings.	Distances	Latitudes.		De-partures.		Corrected Latitudes.		Corrected De-partures.	
			N +	S —	E +	W —	N +	S —	E +	W —
1	N 31¼° W	10.40 ch.	8.87			5.43	8.86			5.44
2	N 62° E	9.20 ch.	4.32		8.13		4.31		8.12	
3	S 36° E	7.60 ch.		6.15	4.47			6.15	4.46	
4	S 45¼° W	10.00 ch.		7.01		7.13		7.02		7.14
		37.20	13.19	13.16	12.60	12.56	13.17	13.17	12.58	12.58

Difference between N and S latitudes = .03 chain = 3 links.

Difference between E and W departures = .04 chain = 4 links.

Corrections for Latitudes.	Corrections for Departures.
37.20 : 10.40 :: 3 : 1 link	37.20 : 10.40 :: 4 : 1 link
37.20 : 9.20 :: 3 : 1 link	37.20 : 9.20 :: 4 : 1 link
37.20 : 7.60 :: 3 : 0 link	37.20 : 7.60 :: 4 : 1 link
37.20 : 10.00 :: 3 : 1 link	37.20 : 10.00 :: 4 : 1 link

Taking the first proportion, we have 37.20 ch., the sum of all the distances : 10.40 ch., the first distance :: 3 links, the total error : 1 link, the correction for the first distance. The latitude of the first course, viz., 8.87 ch., is north, and as the sum of the north latitudes is the greater, we *subtract* the correction leaving 8.86 chains. The correction for the latitudes of the second course is 1 link, and is likewise subtracted. The correction for the third course is less than 1 link, and is ignored. The correction for the latitude of the fourth course is 1 link, and as the sum of the south latitudes is less than the north latitudes we *add* the correction. We place the corrected latitudes in the eighth and ninth columns. In a similar manner we correct the departures, as shown in the above proportions, placing the corrected departures in the tenth and eleventh columns.

(**738**) We rule three columns as shown below, the first column for stations, the second for total latitudes from Sta. 2, and the third for total departures from Sta. 2. Station 2 being a point only, its latitude and departure are 0. The latitude of the second course, i. e., from Sta. 2 to Sta. 3, is + 4.31 chains, and the departure + 8.12 chains. These distances we place opposite Sta. 3, in their proper columns. The latitude of the third course, i. e., from Sta. 3 to Sta. 4, is − 6.15 chains, and the departure + 4.46 chains. Therefore, the total latitude from Sta. 2 is the sum of + 4.31 and − 6.15, which is − 1.84 chains. The total departure from Sta. 2 is the sum of + 8.12 and + 4.46, which is + 12.58 chains. These totals we place opposite Sta. 4 in their proper columns. The latitude of the fourth course, i. e., from Sta. 4 to Sta. 1, is − 7.02 chains, and the departure − 7.14 chains. These quantities we add with their proper signs to those previously obtained, which give us the total latitudes and departures from the initial Sta. 2, and we have, for the total latitude of Sta. 1, the sum of − 1.84 and − 7.02, which is − 8.86 chains, and for the total departure the sum of + 12.58 and − 7.14, which is + 5.44 chains. The latitude of the first course, i. e., from Sta. 1 to Sta. 2, is + 8.86 chains, and the departure − 5.44 chains. These quantities we add with their proper signs to those already obtained, giving us the total latitudes and departures from Sta. 2, and we have for the total latitude of Sta. 2 the sum of − 8.86 and + 8.86, which is 0; and for the total departure, the sum of + 5.44 and − 5.44, which is 0. The latitude and departure of Sta. 2 coming out equal to 0, proves the work to be correct. (See Art. **1332.**) A plat of this survey made from total latitudes and departures from Sta. 2 is given in Fig. 71.

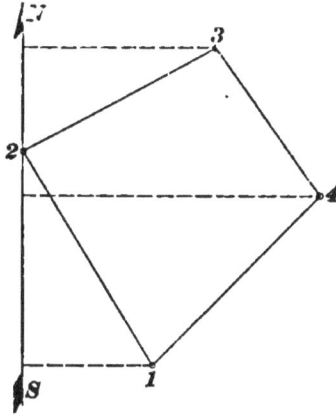

FIG. 71.

Through Sta. 2 draw a meridian *N S*. Lay off on this meridian above Sta. 2 the total latitude at Sta. 3, viz., + 4.31 chains, a north latitude, and at its extremity erect a right perpendicular to the meridian, and upon this perpendicular scale off the total departure of Sta. 3, viz., + 8.12 chains, an east departure locating Sta. 3. A line joining Stations 2 and 3 will have the direction and length of the second course. For Sta. 4 we have a total latitude of − 1.84 chains, a south latitude which we scale off on the meridian below Sta. 2. The total departure of this station is + 12.58 chains, which we lay off on a right perpendicular to the meridian, locating Sta. 4. A line joining Stations 3 and 4 gives the direction and length of the third course.

Stations.	Total Latitudes from Station 2.	Total Departures from Station 2.
2	0.00 ch.	0.00 ch.
3	+ 4.31 ch.	+ 8.12 ch.
4	− 1.84 ch.	+ 12.58 ch.
1	− 8.86 ch.	+ 5.44 ch.
2	0.00 ch.	0.00 ch.

The total latitude of Sta. 1 is − 8.86 chains, a south lati-tude, which we scale off on the meridian below Sta. 2. The total departure of Sta. 1 is + 5.44 chains, an east departure, which we scale off on a right perpendicular to the meridian, locating Sta. 1. A line joining Stations 4 and 1 will have the direction and length of the fourth course. The total latitude and departure of Sta. 2 being 0, a line joining Sta. 1 with Sta. 2 will have the direction and length of the first course, and the resulting figure 2, 3, 4, 1 is the required plat of the survey.

(739) See Art. **1334.**

(740) See Art. **1335.**

(741) See Art. **1336** and Fig. 323..

(742) See Art. **1336.**

(743) The signs of the latitude always determine the character of the products, north or + latitudes giving north products and south or − latitudes giving south products. See Examples 746 and 747.

(744) See Art. **1338.**

(745) Rule twelve columns with headings as shown in the following diagram, calculate the latitudes and departures, placing them in proper order. Balance them, writing the corrected latitudes directly above the original latitudes, which are crossed out. Calculate the double longitudes from Sta. 2 by rule given in Art. **1335,** and place them with their corresponding stations as shown in columns 11 and 12. Place the double longitudes in regular order in column 8. Multiply the double longitude of each course by the cor-rected latitude of that course, placing the products in col-umn 9 or 10, according as the products are north or south. Add the columns of double areas, subtracting the less from the greater and divide the remainder by 2. In this example the area is given in square chains, which we reduce to acres, roods, and poles as follows: Divide the sq. chains by 10, reducing to acres. Multiply the decimal part successively by 4 and 40, reducing to roods and poles.

Stations.	Bearings.	Distances.	Latitudes. N+	Latitudes. S−	Departures. E+	Departures. W−	Double Longitudes. +	Double Longitudes. −	Double Areas. N+	Double Areas. S−	Stations.	D.L.
1	S 21° W	12.41 ch.		11.62 / 11.59		4.44	+ 4.44			51.5928	2	+ 5.82 D.L.
												+ 5.82
												+ 1.72
2	N 83¼° E	5.86 ch.	0.69 / 8.05		5.82		+ 5.82		4.0158		3	+ 13.36 D.L.
												+ 1.72
												− 3.10
3	N 12° E	8.25 ch.	8.07 / 2.88		1.72		+ 13.36		107.5480		4	+ 11.98 D.L.
												− 3.10
												− 4.44
4	N 47° W	4.24 ch.	2.89			3.10	+ 11.98		34.5024		1	+ 4.44 D.L.
	30.76 ch.		11.65 11.59	11.62 11.62	7.54 7.54	7.54 7.54			146.0662			
									51.5928			

Content = 4 A. 2 R. 35.8 P.

$$2\,)\,94.4734$$
$$10\,)\,47.2367 \text{ sq. ch.}$$
$$4.72367$$
$$4$$
$$2.89468$$
$$40$$
$$35.78720$$

In this example, the meridian from which we calculate the double longitudes passes through Sta. 2, the extreme westerly one.

(746)

Stations	Bearings	Distances	Latitudes N+	Latitudes S-	Departures E+	Departures W-	Double Longitudes	Double Areas N+	Double Areas S-	Stations	D. L.
1	S 21¼° W	17.62 ch.		16.41 / 16.42		6.40 / 6.36	− 6.40		105.0240	1	− 6.40 D. L. / − 6.40 / − 5.60
2	S 34° W	10.00 ch.		8.29		5.60 / 5.59 / 11.76 / 11.73	− 18.40		152.5360	2	− 18.40 D. L. / − 5.60 / − 11.76
3	N 56° W	14.15 ch.	7.92 / 7.91				− 35.76	283.2192		3	− 35.76 D. L. / − 11.70 / + 5.45
4	N 34° E	9.76 ch.	8.09		5.45 / 5.40		− 42.07	340.3463		4	+ 42.07 D. L. / + 5.45 / + 2.12
5	N 67° E	2.30 ch.	0.90		2.12		− 34.50	31.0500		5	− 34.50 D. L. / + 2.12 / + 2.75
6	N 23° E	7.03 ch.	6.47		2.75		− 29.63	191.7061		6	− 29.63 D. L. / + 2.75 / + 1.40
7	N 18¼° E	4.43 ch.	4.20		1.40		− 25.48	107.0160		7	− 25.48 D. L. / + 1.40 / + 12.04
8	S 76½° E	12.41 ch.		2.88 / 2.89	12.04 / 12.09		− 12.04		34.6752	8	− 12.04 D. L.
		77.70 ch.	27.27 / 27.58	27.60 / 27.58	23.79 / 23.76	23.70 / 23.76		953.3376 / 292.2352	299.2352		

```
Content = 33 A.  0 R.  8.8 P.
                               953.3376
                               292.2352
                            2)661.1024
                          10)330.5512 sq. ch.
                              33.05512
                                     4
                               .22048
                                    40
                              8.81920
```

In this example, the meridian from which we calculate the double longitudes passes through Sta. 1, the extreme easterly one.

(747)

Sta-tions	Bearings	Distances. Chains.	Latitudes N+	Latitudes S−	Departures E+	Departures W−	Double Longi-tudes	Double Areas N+	Double Areas S−	Stations	D. L.
1	N 18¼° E	1.93	1.83		0.62			2.0130		4	+ 0.91 D.L. + 0.91 + 1.19
2	N 9° W	1.29	1.27			0.20	+ 1.10	1.9304		5	+ 3.01 D.L. + 1.19 + 0.28
3	N 14° W	2.71	2.63			0.66	+ 1.52	1.7358		6	+ 4.48 D.L. + 0.28 + 0.72
4	N 74° E	0.95	0.26				+ 0.66	0.2306		7	+ 5.48 D.L. + 0.72 + 0.72
5	S 48½° E	1.50		1.05	0.91		+ 0.91		3.1605	8	+ 5.71 D.L. + 0.72 − 0.49
6	S 14½° E	1.14		1.10	1.19		+ 3.01		4.9280	9	+ 5.10 D.L. − 0.12 − 0.51
7	S 19½° E	2.15		2.03	0.28		+ 4.48		11.1244	10	+ 4.47 D.L. − 0.51 − 0.68
8	S 23½° W	1.22		1.12	0.72		+ 5.48		6.3952	11	+ 3.28 D.L. − 0.08 − 1.00
9	S 5° W	1.40		1.39		0.49	+ 5.71		7.0890	12	+ 1.54 E.L. − 1.06 + 0.62
10	S 30° W	1.02		0.88		0.12	+ 5.10		3.9336	1	+ 1.10 D.L. + 0.62 + 0.20
11	S 81½° W	0.69		0.10		0.51	+ 4.47		0.3280	2	+ 1.52 D.L. − 0.20 − 0.60
12	N 32¼° W	1.98	1.67			0.68	+ 3.28	2.5718		3	+ 0.66 D.L.
						1.06	+ 1.54				
			7.66	7.67	3.72	3.72		8.4876	36.9587		

```
                              36.9587
                               8.4876
                           2)28.4711
           Sq. chains  10)14.2355
                           1.42355
                                 4
                           1.59420
                                40
                          27.76800
```

Content, 1 A.   1 R.   27¼ P.

In this example, the meridian from which we calculate the double longitudes passes through Sta. 4, the extreme westerly one.

(**748**) See Art. **1339.**

(**749**) See Art. **1340** and Fig. 326.

(**750**) See Art. **1341** and Fig. 327.

(**751**) See Art. **1341.**

(**752**) See Art. **1342.**

(**753**) See Art. **1343** and Fig. 327.

(**754**) See Art. **1344.**

(**755**) See Art. **1343.**

RAILROAD LOCATION.

(QUESTIONS 756–812.)

(**756**) See Art. **1392.**

(**757**) See Art. **1393.**

(**758**) See Art. **1393.**

(**759**) See Art. **1394.**

(**760**) This question is asked in order to elicit *individual* judgment.

(**761**) See Art. **1395.**

(**762**) See Art. **1396.**

(**763**) See Art. **1397.**

(**764**) See Art. **1398.**

(**765**) See Art. **1399.**

(**766**) See Art. **1399.**

(**767**) See Art. **1400.**

(**768**) See Art. **1401.**

(**769**) See Art. **1402.**

(**770**) See Art. **1403.**

(**771**) See Art. **1404.**

(**772**) See Art. **1405.**

(**773**) See Arts. **1406** and **1407.**

(**774**) See Art. **1408.**

(775) See Art. **1409.**

(776) See Art. **1409.**

(777) See Arts. **1410** and **1412.**

(778) See Art. **1413.**

(779) See Art. **1414.**

(780) See Art. **1415.**

(781) See Art. **1416.**

(782) See Art. **1416.**

(783) See Art. **1417.**

(784) See Art. **1418.**

(785) See Art. **1419.**

(786) See Art. **1420.**

(787) This is an example under Problem I. Art. **1422.** The distance which the P. C. must be moved backwards is equal to 26 ft., the distance between the parallel tangents, divided by the sin of 32° 30′, the intersection angle of the curve. Sin 32° 30′ = .5373; $\frac{26 \text{ ft.}}{.5373} = 48.39$ ft.

(788) This question also comes under Problem I, Art. **1422.** We must move the P. C. forwards a distance equal to 13.4 divided by sin 41° 20′. Sin 41° 20′=.66044. $\frac{13.4 \text{ ft.}}{.66044} =$ 20.29 ft.

(789) This question comes under Problem II, Case 2, Art. **1424.** As we must move the P. C. C. backwards, we must increase the angle of the second curve. This will diminish the cos of the angle of the second curve, and D, the distance between the tangents, will be *negative*. Accordingly, we use formula **100,** $\cos y = \frac{(R - r) \cos x - D}{R - r}$ (see Art. **1424**), in which $x = 34°$ 20′, the angle of the

second curve; $R = 955.37$ ft., the radius of the 6° curve, and $r = 637.27$ ft., the radius of 9° curve, and $D = 26.4$ ft. Substituting these values in the foregoing formula, we have

$$\cos y = \frac{(955.37 - 637.27) \times .82577 - 26.4}{955.37 - 637.27} =$$

0.7428, the cos of 42° 02'.

As the given angle of the second curve is 34° 20', and the required angle 42° 02', the difference, viz., 7° 42', we must deduct from the first curve. The distance which we must retreat on the first curve we determine by dividing the angle 7° 42' by 6, the degree of the first curve. The quotient will be the required distance in stations. Reducing 7° 42' to the decimal of a degree, we have 7.7°. $\dfrac{7.7}{6} = 1.2833$ stations $= 128.33$ ft.

(790) This question comes under Problem II, Case 1, Art. 1423, but the tangent falls within instead of without the required tangent. Consequently, we must advance the P. C. C., which will diminish the angle of the second curve and, consequently, increase its cos. D, the distance between the given tangents, will, therefore, be *positive*, giving us formula 99, viz., $\cos y = \dfrac{(R - r) \cos x + D}{R - r}$ (see Art. 1423), in which $x = 36° 40'$, the angle of the second curve; $R = 1,910.08$ ft., the radius of a 3° curve; $r = 819.02$ ft., the radius of a 7° curve, and $D = 32.4$ ft., the distance between the tangents. Substituting these values in the given formula, we have

$$\cos y = \frac{(1,910.08 - 819.02) \times .80212 + 32.4}{1,910.08 - 819.02} =$$

.83181, the cos of 33° 43'.

The given angle of the second curve is 36° 40', and the required angle is 33° 43'. The difference, 2° 57', we must deduct from the second curve and add it to the first curve. To determine the number of feet which we must add to the first curve, we divide the angle 2° 57' by 3, the degree of the first curve. The quotient will be the distance in full stations. Reducing 2° 57' to decimal form of a degree, we have 2.95°. $\dfrac{2.95}{3} = 0.9833$ station $= 98.33$ ft., the distance which the P. C. C. must be advanced.

(**791**) This question comes under Problem II, Case 1. (See Art. **1423**.) In this case we must increase the angle of the second curve, and, consequently, diminish the length of the cos. Hence, D, the distance between the given tangents, will be negative, and the formula will read

$$\cos y = \dfrac{(R - r)\cos x - D}{R - r} \text{ (see formula } \mathbf{100}, \text{ Art. } \mathbf{1423}),$$

in which $x = 37° 10'$, the angle of the second curve; $R = 1,432.69$ ft., the radius of a 4° curve; $r = 955.37$ ft., the radius of a 6° curve, and $D = 56$ ft., the distance between the given tangents. Substituting these values in the given formula, we have

$$\cos y = \dfrac{(1,432.69 - 955.37) \times .79688 - 56}{1,432.69 - 955.37} = .67956 = \cos 47° 11'.$$

The given angle of the second curve is 37° 10', and the required angle, 47° 11'. The difference, viz., 10° 01', we must add to the second curve and deduct from the first curve. The distance backwards which we must move the P. C. C. we obtain by dividing the angle 10° 01' by 4, the degree of the first curve. The quotient gives the distance in full stations. 10° 01' in decimal form is 10.016°. $\dfrac{10.016}{4} = 2.504$ stations $= 250.4$ ft.

(**792**) This question comes under Problem II, Case 2, Art. **1424**. Here we must reduce the angle of the second curve, and, consequently, increase the length of its cos. Hence, D, the distance between the given tangents, is positive, and our formula is

$$\cos y = \dfrac{(R - r)\cos x + D}{R - r},$$

in which $x = 28° 40'$, the given angle of the second curve; $R = 1,910.08$ ft., the radius of a 2° curve; $r = 716.78$ ft., the radius of an 8° curve, and $D = 25.4$ ft., the distance between the given tangents. Substituting these values in the above formula, we have

$$\cos y = \frac{(1,910.08 - 716.78) \times .87743 + 25.4}{1,910.08 - 716.78} = .89872 = \cos 26°00'.$$

The given angle of the second curve is $28° 40'$, and the required angle is $26° 00'$. The difference, $2° 40'$, we must deduct from the second curve and add to the first, i. e., we must advance the P. C. C. The number of feet which we advance the P. C. C. we determine by dividing the angle $2° 40'$ by 8, the degree of the first curve; the quotient gives the distance in stations. Reducing $2° 40'$ to the decimal of a degree, we have $2.6667°$. $\frac{2.6667}{8} = .3333$ station $= 33.33$ ft.

(**793**) This question is also under Problem II, Case 2. (See Art. **1424**.) Here we increase the angle of the second curve, and, consequently, diminish the cos, and D the distance between the tangents is negative. We use formula **100**, $\cos y = \frac{(R - r) \cos x - D}{R - r}$, in which $x = 36° 15'$, $R = 1,432.69$ ft., $r = 637.27$ ft., and $D = 33$ ft. Substituting these values in the given formula, we have

$$\cos y = \frac{(1,432.69 - 637.27) \times .80644 - 33}{1,432.69 - 637.27} = .76495 \cos 40° 06'.$$

The given angle of the second curve is $36° 15'$ and the required angle $40° 06'$. The difference, viz., $3° 51'$, we add to the second curve and deduct from the first. We must, therefore, place the P. C. C. back of the given P. C. C., and this distance we find by dividing the angle $3° 51'$ by 9, the degree of the first curve. Reducing $3° 51'$ to decimal form, we have $3.85°$. $\frac{3.85}{9} = .4278$ station $= 42.78$ ft.

(**794**) This question comes under Problem III, Art. **1425**. The radius of a 7° curve is 819.02 ft. The radius of the parallel curve will be $819.02 - 100 = 719.02$ ft. The chords on the $7°$ curve are each 100 ft., and we obtain the length of the parallel chords from the following proportion:

$$819.02 : 719.02 :: 100 \text{ ft.} : \text{the required chord.}$$

Whence, we have required chord $= \dfrac{719.02 \times 100}{819.02} = 87.79$ ft.

(**795**) This question comes under Problem IV, Art. **1426.**

$$\frac{34° \ 20'}{2} = 17° \ 10'; \ \frac{41° \ 30'}{2} = 20° \ 45'.$$

The distance between intersection points is 1,011 ft. From Art. **1426,** we have

(tan $17° \ 10' +$ tan $20° \ 45'$) : tan $17° \ 10' :: 1,011$: the tangent distance of the first curve.

Whence, $(.30891 + .37887) : .30891 :: 1,011$ ft. : the tangent distance.

Whence, tangent distance of the first curve $= \dfrac{312.308}{.68778} = $ 454.08 ft.

Substituting known values in formula **91,** $T = R \tan \frac{1}{2} I$ (see Art. **1251**), we have $454.08 = R \times .30891$; whence, $R = \dfrac{454.08}{.30891} = 1,469.94$ ft. Dividing 5,730 ft., the radius of a $1°$ curve, by 1,469.94, the length of the required radius, the quotient 3.899 is the degree of the required curve. Reducing the decimal to minutes, we have the degree of the required curve $= 3° \ 53.9'$.

(**796**) This question also comes under Problem IV, Art. **1426.** We have

$$\frac{20° \ 14'}{2} = 10° \ 07'; \ \frac{41° \ 08'}{2} = 20° \ 34'.$$

The distance between intersection points is 816 ft. From Art. **1426,** we have

(tan $10° \ 07' +$ tan $20° \ 34'$) : tan $10° \ 07' :: 816$ ft. : the tangent distance of the first curve.

Whence, we have $(.17843 + .37521, : .17843 :: 816 :$ tangent distance.

Whence, tangent distance $= \dfrac{145.5989}{.55364} = 262.985$ ft.

Substituting known values in formula **91**, $T = R \tan \frac{1}{2} I$ (see Art. **1251**), we have $262.985 = R \times .17843$; whence, $R = \dfrac{262.985}{.17843} = 1{,}473.88$ ft. $=$ the radius of a $3°$ $53.3'$ curve.

(**797**) This question comes under Problem V, Art. **1427.** The required radius is equal to 470 ft. divided by $(\tan \frac{1}{2} \, 28° \, 40' + \tan \frac{1}{2} \, 30° \, 16')$. $\dfrac{28° \, 40'}{2} = 14° \, 20'$; $\tan$ $14° \, 20' = .25552$. $\dfrac{30° \, 16'}{2} = 15° \, 08'$; $\tan 15° \, 08' = .27044$. The sum of these tangents is $.52596$, and we have $R = \dfrac{470 \text{ ft.}}{.52596} = 893.64$ ft. Dividing 5,730 ft., the radius of a $1°$ curve, by the radius 893.64 ft., the quotient is the degree of the required curve. $\dfrac{5{,}730}{893.64} = 6.412° = 6° \, 24.7'$ curve.

(**798**) This question also comes under Problem V, Art. **1427.** The required radius $R = \dfrac{516 \text{ ft.}}{\tan \frac{1}{2} 32° \, 50' + \tan \frac{1}{2} 41° \, 20'}$. $\dfrac{32° \, 50'}{2} = 16° \, 25'$. $\tan 16° \, 25' = .29463$. $\dfrac{41° \, 20'}{2} = 20° \, 40'$. $\tan 20° \, 40' = .3772$. $.29463 + .37720 = .67183$. Substituting this value in the above equation, we have $R = \dfrac{516}{.67183} = 768.05$ ft., the radius of a $7° \, 27.6'$ curve.

(**799**) This question comes under Problem VII, Art. **1429.** The required distance across the stream $= \dfrac{7.3 \times 100}{1.745} = 418.3$ ft.

(**800**) See Art. **1433.**

(801)   See Art. **1434.**

(802)   See Art. **1435.**

(803)   The usual compensations for curvature are from .03 ft. to .05 ft. per degree.

(804)   As the elevation of grade at Sta. 20 is 118.5 ft., and that at Sta. 40 is 142.5 ft., the total actual rise between those stations is the difference between 142.5 and 118.5, which is 24 ft.   Hence, the average grade between those points is $\frac{24}{20} = 1.2$ ft. per station.   As the resistance owing to the curvature is equivalent to an increase in grade of .03 ft. per each degree, the total increase in grade owing to curvature is equal to .03 ft. multiplied by 78, the total number of degrees of curvature between Sta. 20 and Sta. 40.   78 × .03 = 2.34 ft.   This amount we add to 24 ft., the total actual rise between the given stations, making a total *theoretical* rise of 26.34 ft.   Dividing 26.34 by 20, we obtain for the tangents on this portion of the line an ascending grade of 1.317 ft. per station.   Hence, the grade between Sta. 20 and Sta. 24 + 50 is + 1.317 ft. per station.   The distance between Sta. 20 and Sta. 24 + 50 is 450 ft. = 4.5 stations, and the total rise between these stations is 1.317 ft. × 4.5 = 5.9265 ft., which we add to 118.5 ft., the elevation of grade at Sta. 20, giving 124.4265 ft. for the elevation of grade at Sta. 24 + 50.   The first curve is 10°, which is equivalent to a grade of .03 ft. × 10 = 0.3 ft. per station, which we subtract from 1.317, the grade for tangents.   The difference, 1.017, is the grade on the 10° curve, the length of which is 420 ft. = 4.2 stations.   Multiplying 1.017 ft., the grade on the 10 curve, by 4.2, we have 4.2714 ft. as the total rise on that curve, the P. T. of which is Sta. 28 + 70.   Adding 4.2714 ft. to 124.4265 ft., we have 128.6979 ft., the elevation of grade at Sta. 28 + 70.   The line between Sta. 28 + 70 and Sta. 31 + 80 being tangent, has a grade of 1.317 ft.   The distance between these stations is 310 ft. = 3.1 stations, and the total rise between the stations is 1.317 × 3.1 =

4.0827 ft., which we add to 128.6979 ft., the elevation of
grade at Sta. 28 + 70, giving 132.7806 ft. for the elevation
of grade at Sta. 31 + 80.   Here we commence an 8° curve
for 450 ft. = 4.5 stations.   The compensation in grade for
an 8° curve is .03 ft. × 8 = 0.24 ft. per station.   Hence, the
grade for that curve is 1.317 ft. − 0.24 ft. = + 1.077 ft. per
station, and the total rise on the 8° curve is 1.077 ft. × 4.5 =
4.8465 ft., which we add to 132.7806, the elevation of
grade at Sta. 31 + 80, giving 137.6271 ft. for the elevation
of grade at Sta. 36 + 30, the P. T. of the 8° curve.   The
line between Sta. 36 + 30 and Sta. 40 is a tangent, and has
a grade of + 1.317 ft. per station.   The distance between
these stations is 370 ft. = 3.7 stations, and the total rise is
1.317 × 3.7 = 4.8729 ft., which, added to 137.6271 ft., the
elevation of grade at Sta. 36 + 30, gives 142.5 ft. for the
elevation of grade at Sta. 40.

(805)   See Art. **1439.**

(806)   In this question, $g = + 1.0$ ft., $g' = − 0.8$ ft.
and $n = 3$.   Substituting these values in formula **101,**
$a = \dfrac{g - g'}{4n}$ (see Art. **1440**), we have $a = \dfrac{1.0 - (- 0.8)}{12} =$
$\dfrac{1.8}{12} = 0.15$ ft.   The successive grades or additions for the
6 stations of the vertical curve are the following: $g - a$,
$g - 3a, g - 5a, g - 7a, g - 9a, g - 11a$.   Substituting known
values of g and a, we have for the successive grades:

				Heights of Curve Above Starting Point.
1.	$g -$	$a = 1.0$ ft. − 0.15 ft. =	0.85 ft.	....0.85 ft.
2.	$g -$	$3a = 1.0$ ft. − 0.45 ft. =	0.55 ft.	....1.40 ft.
3.	$g -$	$5a = 1.0$ ft. − 0.75 ft. =	0.25 ft.	....1.65 ft.
4.	$g -$	$7a = 1.0$ ft. − 1.05 ft. =	− 0.05 ft.	....1.60 ft.
5.	$g -$	$9a = 1.0$ ft. − 1.35 ft. =	− 0.35 ft.	....1.25 ft.
6.	$g -$	$11a = 1.0$ ft. − 1.65 ft. =	− 0.65 ft.	....0.60 ft.

As the elevation of the grade at the starting point of the

curve (which we will call Sta. 0) is 110 ft., the elevations ot the grades for all the stations of the curve are the following:

Stations.	Elevation of Grade.
0	110.00
1	110.85
2	111.40
3	111.65
4	111.60
5	111.25
6	110.60

(**807**)　In Fig. 72, A C is an ascending grade of 1 per cent., and C B is a descending grade of 0.8 per cent., which

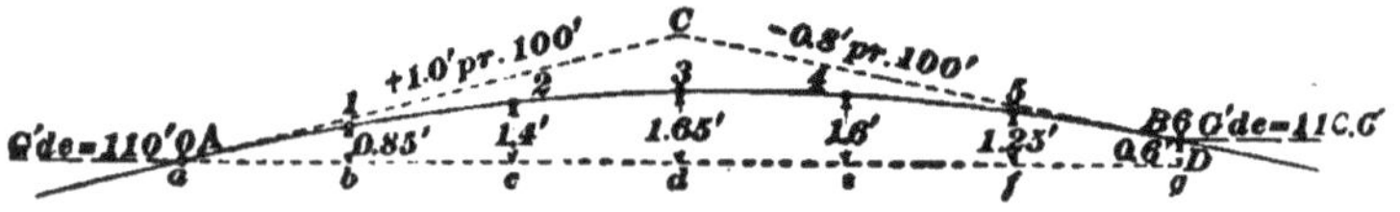

FIG. 72.

are the grades specified in Question 806.　To draw these grade lines, first draw a horizontal line A D 6 in. long, which will include both grade lines to a scale of 100 ft. to the inch.　Divide this line into six equal parts, with the letters a, b, c, etc., at the points of division.　Now, d being 300 ft. from A, the height of the original grade line above d is the rate of grade.　$g = 1.0$ ft. $\times 3 = 3$ ft., which distance we scale off above d to a vertical scale of 5 ft. to the inch, locating the point C.　The grade of C B is -0.8 ft. per 100 ft.　Consequently, B, which is 300 ft. from C, is 0.8 ft. $\times 3 = 2.4$ ft. below C, and the height of B above A D is equal to 3.0 ft. $- 2.4$ ft. $= 0.6$ ft.　We scale off above D the distance 0.6 ft., locating B.　Joining A C and B C, we have the original grade lines, which are to be united by a vertical curve.　Now, the elevation of the grade at Sta. 0 is 110 ft. The line A D has the same elevation.　We have already determined the heights of the curve above this line at the several stations on the curve.　Accordingly, we lay off these distances, viz., at b, corresponding to Sta. 1, 0.85 ft.; at c, Sta. 2, 1.4 ft.; at d, Sta. 3, 1.65 ft., etc., marking each point so determined by a small circle. The curved line joining these points is the vertical curve required.

(808) See Art. **1443.**

(809) See Art. **1444.**

(810) See Art. **1446.**

(811) See Art. **1447.**

(812) See Art. **1451.**

RAILROAD CONSTRUCTION.

---

(813) See Art. **1454.**

(814) See Art. **1455.**

(815) The slope given to embankment is $1\frac{1}{2}$ horizontal to 1 vertical, and the slope usually given to cuttings is 1 horizontal to 1 vertical. (See Art. **1457** and Figs. 380 and 381.)

(816) The height of the instrument being 127.4 feet and the elevation of the grade 140 feet, the instrument is below grade to an amount equal to the difference between 140 and 127.4 feet, which is 12.6 feet. The rod reading for the right slope being 9.2 feet, the surface of the ground is 9.2 feet below the instrument, which we have already shown to be 12.6 feet below grade. Hence, the distance which the surface of the ground is below grade is the sum of 12.6 and 9.2 which is 21.8 feet, which we describe as a *fill* of 21.8 feet. Ans. The side distance, i. e., the distance at which we must place the slope stake from the center line, is $1\frac{1}{2}$ times 21.8 feet, the amount of the fill, plus $\frac{1}{2}$ the width of the roadway, or 8 feet. Therefore, side distance $= \dfrac{3 \times 21.8}{2} + 8 =$ 40.7 feet. Ans.

(817) The height of instrument H. I. being 96.4 feet, and the rod reading at the surface of the ground 4.7 feet, the elevation of the surface of the ground is $96.4 - 4.7 = 91.7$ feet. As the elevation of grade is 78.0 feet, the amount of cutting is the difference between 91.7 and $78.0 = 13.7$ feet. Ans.

(818)   The elevation of the surface of the ground is the height of instrument minus the rod reading; i. e., 96.4 − 8.8 = 87.6 feet.   The elevation of grade is 78.0 feet; hence, the cutting is 87.6 − 78.0 = 9.6 feet.   The slope of the cutting is 1 foot horizontal to 1 foot vertical; hence, from the foot of the slope to its outer edge is 9.6 feet.   Ans.   To this we add one-half the width of the roadway, or 9 feet, giving foi the side distance 9.6 + 9 = 18.6 feet.   Ans.

(819)   See Art. **1459.**

(820)   See Art. **1460.**

(821)   See Art. **1461.**

(822)   We substitute the given quantities in formula **102** $A = C\sqrt{M}$ (see Art. **1461**), in which A = the area of the culvert opening in square feet; C the variable coefficient, and M the area of the given water shed in acres.   We accordingly have $A = 1.8\sqrt{400} = 36$ square feet.   Ans.

(823)   Applying rule I, Art. **1462,** we obtain the distance from the center line to the face of the culvert as follows :   To the height of the side wall, viz., 4 feet, we add the thickness of the covering flags and the height of the parapet—each 1 foot, making the total height of the top of parapet 4 + 2 = 6 feet.   28 feet, the height of the embankment at the center line, minus 6 feet = 22 feet.   With this as the height of the embankment, we calculate the side distance as in setting slope stakes $1\frac{1}{2}$ times 22 feet = 33 feet. One-half the width of roadway is $\frac{16}{2} = 8$ feet.   33 + 8 = 41 feet.

To this distance we add 18 inches, making 42 feet 6 inches. As the embankment is more than 10 feet in height, we add to this side distance 1 inch for each foot in height above the parapet, i. e., 22 inches = 1 foot 10 inches.   Adding 1 foot 10 inches to 42 feet 6 inches, we have for the total distance from the center line to the face of the culvert 44 feet 4 inches.   Ans.

We find the length of the wing walls by applying rule II,

Art. **1462.** The height of top of the covering flags is 5 feet. 1½ times 5 feet = 7.5 feet. Adding 2 feet, we have length of the wing walls, i. e., the distance from inside face of the side walls to the end of the wing walls, 7.5 + 2 = 9 feet 6 inches. Ans.

(**824**) The span of a box culvert should not exceed 3 feet. When a larger opening is required, a double box culvert with a division wall 2 feet in thickness is substituted.

(**825**) See Art. **1462.**

(**826**) See Art. **1462.**

(**827**) See Art. **1463.**

(**828**) See Art. **1465.**

(**829**) The thickness of the base should be $\frac{4}{10}$ of the height. The height is 16 feet and the thickness of the base should be $\frac{4}{10}$ of 16 feet, or 6.4 feet. Ans.

(**830**) See Art. **1468.**

(**831**) See Art. **1468.**

(**832**) Trautwine's formula for finding the depth of keystone (see formula **103**, Art. **1469**), is as follows:

$$\text{depth of keystone in feet} = \frac{\sqrt{\text{radius of arch} + \frac{1}{2}\text{span}}}{4} + 0.2 \text{ foot.}$$

In this example the arch being semicircular, the radius and half-span are the same. Substituting known values in the given formula, we have

$$\text{depth of keystone} = \frac{\sqrt{15 + 15}}{4} + 0.2 \text{ foot} = 1.57 \text{ feet. Ans.}$$

Applying Rankine's formula (formula **104**), depth of keystone = $\sqrt{.12 \text{ radius}}$, we have depth of keystone = $\sqrt{.12 \times 15} = 1.34$ feet. Ans.

(833) Applying the rule given in Art. **1469,** we find the length of radius is equal to $\dfrac{19^2 + 12^2}{24} = \dfrac{505}{24} = 21.04$ feet.

Ans.

(834) Applying formula **105,** Art. **1470,** we have

$$\text{thickness of abutment at spring line} = \frac{12 \text{ feet}}{5} + \frac{8 \text{ feet}}{10} + 2 \text{ feet} = 5.2 \text{ feet.}$$

Ans.

(835) See Art. **1472.**

(836) See Art. **1471.**

(837) See Art. **1472.**

(838) See Art. **1473.**

(839) See Art. **1473.**

(840) See Art. **1474.**

(841) See Art. **1475.**

(842) See Art. **1475.**

(843) They should be laid in the radial lines of the arch. See Art. **1477.**

(844) See Art. **1477.**

(845) Until the arch is half built, the effect of the weight of the arch upon the centering is to cause a lifting at the crown. After that point is passed, the effect of the weight is to cause a lifting of the haunches.

(846) 120°. Ans.

(847) See Art. **1480.**

(848) According to the rule given in Art. **1480,** the thickness of the base should be $\frac{4}{10}$ of the vertical height 10 feet $\times \frac{4}{10} = 4$ feet. Ans.

(849) See Art. **1480.**

(850)   The friction of the backing against the wall adds considerably to its stability.

(851)   See Art. **1481.**

(852)   See Art. **1482.**

(853)   See Art. **1483** and Fig. 401.

(854)   See Art. **1484.**

(855)   The line *d c* in Fig. 73 forms an angle of 33° 41′ with the horizontal *d h* and is the natural slope of earth.

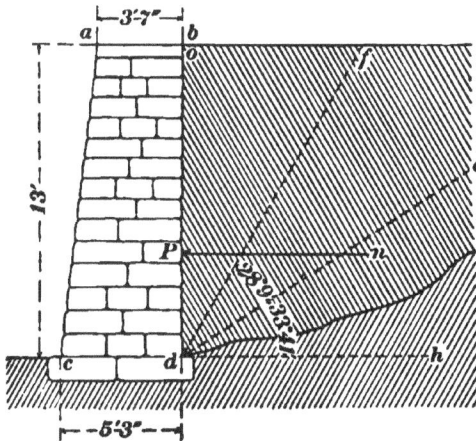

FIG. 73.

The line *d f* bisects the angle *o d c*.   The angle *o d f* is called the *angle*, the slope *d f* is called the *slope*, and the triangular prism *o d f* is called the *prism* of maximum pressure.

(856)   See Art. **1486.**

(857)   See Art. **1486.**

(858)   From Fig. 73, given above, we have, by applying formula **106,** Art. **1486,**

$$\text{perpendicular pressure } n \; P = \frac{\text{the weight of the triangle of earth } odf \times of}{\text{the vertical height } od}.$$

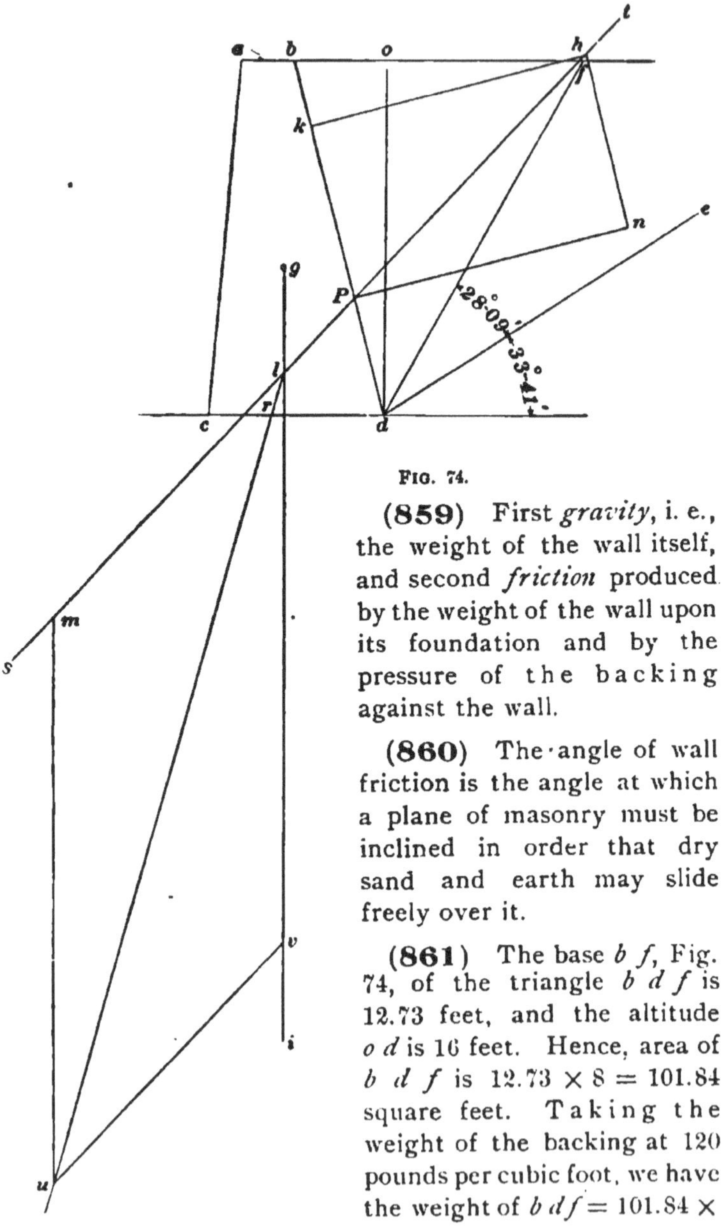

FIG. 74.

(859) First *gravity*, i. e., the weight of the wall itself, and second *friction* produced by the weight of the wall upon its foundation and by the pressure of the backing against the wall.

(860) The angle of wall friction is the angle at which a plane of masonry must be inclined in order that dry sand and earth may slide freely over it.

(861) The base $b\,f$, Fig. 74, of the triangle $b\,d\,f$ is 12.73 feet, and the altitude $o\,d$ is 16 feet. Hence, area of $b\,d\,f$ is $12.73 \times 8 = 101.84$ square feet. Taking the weight of the backing at 120 pounds per cubic foot, we have the weight of $b\,d\,f = 101.84 \times$

120 = 12,221 pounds. Multiplying this weight by $f=$ 8.56 feet, we have $12,221 \times 8.56 = 104,612$ pounds, which, divided by $o\,d$, 16 feet, $= 6,538$ pounds $=$ the pressure of the backing. This pressure to a scale of 4,000 pounds to the inch equals 1.63 inches. P, the center of pressure, is at $\frac{1}{3}$ of the height of $b\,d$ measured from d. At P erect a perpendicular to the back of the wall. Lay off on this perpendicular the distance $Pn = 1.63$ in. Draw Pt, making the angle $n\,P\,t = 33° 41' =$ the angle of wall friction. At n draw a perpendicular to Pn, intersecting the line Pt in h. Draw $h\,k$, completing the parallelogram $n\,h\,k\,P$; $h\,n$, or its equal $k\,P$, will represent the friction of the backing against the wall and the diagonal $h\,P$, which, to the same scale, $=$ 7,857 pounds, will be the resultant of the pressure of the backing and the friction. Produce $P\,h$ to s. The section $a\,b\,d\,c$ of the wall is a trapezoid. Its area is 84 square feet, which, multiplied by 154 pounds, the weight of rubble per cubic foot, gives 12,936 pounds, as the weight of the wall, which, to a scale of 4,000 pounds to the inch, $= 3.23$ inches. Find the center of gravity g of the section $a\,b\,d\,c$, as explained in Art. **1488.** Through g draw the vertical line $g\,i$, intersecting the prolongation of $P\,h$ in l. Lay off from l, on $g\,i$, the distance $l\,v$, 3.23 inches $=$ weight of the wall, and on $l\,s$, the distance $l\,m = 1.96$ inches, the length of the resultant $P\,h$; complete the parallelogram $l\,m\,u\,v$. The diagonal $l\,u$ is the resultant of the weight of the wall and the pressure. The distance $c\,r$ from the toe c to the point where the resultant $l\,u$ cuts this base is 3.6 feet, or nearly $\frac{1}{4}$ of the base $c\,d$. This guarantees abundant stability to the retaining wall.

The distance $o\,f$ is obtained as follows: $c\,d = 8$ feet; batter of $c\,a = 1$ inch for each foot of length of $o\,d = 16$ inches; hence, $a\;o = 8$ feet $- 16$ inches $= 6$ feet 8 inches; $a\;b = 2.5$ feet; therefore, $o\,f = 12.73 - (6$ feet 8 inches $- 2.5$ feet$) = 8.56$ feet.

(**862**)   See Art. **1491.**

(**863**)   See Art. **1491** and Figs. 409, 410, and 411.

(864)   See Art. **1492.**

(865)   See Art. **1493.**

(866) $\dfrac{540, \text{ the number of working minutes in a day,}}{1.25 + 4} =$

102.9 trips. $\dfrac{102.9}{14} = 7.35$ cubic yards per man; $\dfrac{\$1.15}{7.35} =$

15.64 cents per cubic yard for wheeling.   One picker serves 5 wheelbarrows; he will accordingly loosen $7.35 \times 5 = 36.75$

cubic yards.   Cost of picking will, therefore, be $\dfrac{\$1.15}{36.75} =$

3.13 cents per cubic yard.   There are 25 men in the gang, one-fifth of whom are pickers.   There are, consequently, 20 wheelers, who together will wheel in one day $7.35 \times 20 = 147$ cubic yards.   One foreman at $2.00 and one water-carrier at 90 cents per day are required for such a gang.   Their combined wages are $2.90.   The cost of superintendence

and water-carrier is, therefore, $\dfrac{\$2.90}{147} = 1.97$ cents per cubic

yard.   Use of tools and wheelbarrows is placed at $\frac{1}{2}$ cent per cubic yard.

Placing items of cost in order, we have

Cost of wheeling.............   15.64 cents per cubic yard.
Cost of picking ....   3.13 cents per cubic yard.
Cost of water-carrier and super-
    intendence ....   1.97 cents per cubic yard.
Use of tools and wheelbarrows.   .50 cents per cubic yard.
                                 ─────
                                 21.24 cents per cubic yard.

(867)   The number of carts loaded by each shoveler

(Art. **1494**), is $\dfrac{420}{5} = 84$. $84 \div 3 = 28$ cubic yards handled

per day by each shoveler.   The cost of shoveling is, there-

fore, $\dfrac{120}{28} = 4.28$ cents per cubic yard.   The number of cart

trips per day is $\dfrac{600}{7 + 4} = 54.54$.   As a cart carries $\frac{1}{3}$ cubic

yard, each cart will in one day carry $\dfrac{54.54}{3} = 18.18$ cubic yards. As cart and driver cost $1.40 per day, the cost of hauling is $\frac{\$1.40}{18.18} = 7.7$ cents per cubic yard. Foreman and water-carrier together cost $3.25. The gang contains 12 carts, which together carry $18.18 \times 12 = 218.2$ cubic yards. $\frac{\$3.25}{218.2} = 1.49$ cents per cubic yard for water-carrier and superintendence. As loosening soil costs 2 cents per cubic yard, dumping and spreading 1 cent per cubic yard, and wear of carts and tools $\frac{1}{2}$ cent per cubic yard, we have the total cost per cubic yard as follows:

Loosening soil................ 2.00 cents per cubic yard.
Shoveling into carts.......... 4.28 cents per cubic yard.
Hauling .... 7.70 cents per cubic yard.
Superintendence and water-car-
  rier .... 1.49 cents per cubic yard.
Wear and tear of carts and tools, 0.50 cents per cubic yard.
Dumping and spreading........ 1.00 cents per cubic yard.

Cost for delivering on the dump, 16.97 cents per cubic yard.

(868) See Art. **1499.**

(869) See Art. **1499.**

(870) See Art. **1500.**

(871) See Art. **1500.**

(872) In carrying the steam a long distance through iron pipes its pressure is greatly reduced by condensation, whereas compressed air may be carried a great distance without suffering any loss in pressure excepting that due to friction and leakage.

RAILROAD CONSTRUCTION.

(873) See Art. **1503.**

(874) See Art. **1504.**

(875) See Art. **1504.**

(876) See Art. **1505** and Figs. 426 and 428.

(877) As 60° F. is assumed as normal temperature, and as the temperature at the time of measuring the line is 94°, we must, in determining the length of the line, make an allowance for expansion due to an increase of temperature equal to the difference between 60° and 94°, or 34°. The allowance per foot per degree, as stated in Art. **1505,** is .0000066 ft., and for 34° the allowance per foot is .0000066 × 34 = .0002244 ft.; for 89.621 ft. the allowance is .0002244 × 89.621 = 0.020 ft. The normal length of the line will, therefore, be 89.621 + .020 = 89.641 ft. Ans.

(878) Let the line $A B$ in Fig. 75 represent the slope distance as measured, viz., 89.72 ft. This line will, together with the difference of elevation between the extremities A and B, viz., 11.44 ft., and the

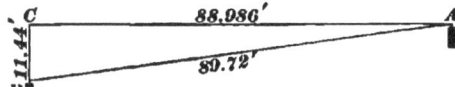

FIG. 75.

required horizontal distance $A C$, form a right-angled triangle, right-angled at C. By rule **1,** Art. **754,** we have $\sin A = \dfrac{11.44}{89.72} = .12751$; whence, $A = 7°\ 20'$. Again, by rule **3,** Art. **754,** $\cos 7°\ 20' = \dfrac{A C}{89.72}$; whence, $.99182 = \dfrac{A C}{89.72}$, and $A C = 89.72 \times .99182 = 88.986$ ft. Ans.

(**879**) See Art. **1508.**

(**880**) See Art. **1511.**

(**881**) See Art. **1511** and Figs. 435, 436, 445, and 446.

(**882**) See Art. **1513.**

(**883**) See Art. **1514.**

(**884**) See Art. **1517.**

(**885**) If the elevation of the grade of the station is 162 ft. and the height of the tunnel section is 24 ft., the elevation of the tunnel roof at that station is $162 + 24 = 186$ ft. The height of instrument is 179.3 ft., and when the roof at the given station is at grade, the rod reading will be $186 - 179.3 = 6.7$ ft. Ans.

(**886**) See Art. **1515.**

(**887**) See Art. **1517** and Fig. 448.

(**888**) See Art. **1520** and Fig. 452.

(**889**) See Art. **1523.**

(**890**) See Art. **1525.**

(**891**) See Art. **1526.**

(**892**) The area of an 18-inch air pipe is $1.5' \times .7854 = 1.77$ sq. ft. At a velocity of 13 ft. per second, the amount of foul air removed from the heading per second is $1.77 \times 13 = 23.01$ cu. ft. And as each cubic foot of foul air removed is replaced by one of pure air, in 1 minute the amount of pure air furnished is $23.01 \times 60 = 1,380.6$ cu. ft. As each man requires 100 cu. ft. of pure air per minute, there will be a supply for as many laborers as 100 is contained times in 1,380.6, which is 13.8, say 14. Ans.

(**893**) See Art. **1528.**

(**894**) See Art. **1533** and Fig. 455.

(**895**) See Art. **1535** and Fig. 457.

(**896**)   See Art. **1536** and Fig. 458.

(**897**)   The height of embankment is 21 ft. ; the culvert opening, 3 ft. in height, and the covering flags and parapet each 1 ft. in height, making the top of parapet 5 ft. above the foundation.   In Fig. 76 the angle $B A L$ is 75° and represents the skew of the culvert.   Drawing the line $A D$ at right angles to the center line $A C$, we have the angle $B A D = 90° - 75° = 15°$.   If the culvert were built at right angles to the center line, the side distance $A D$ from the

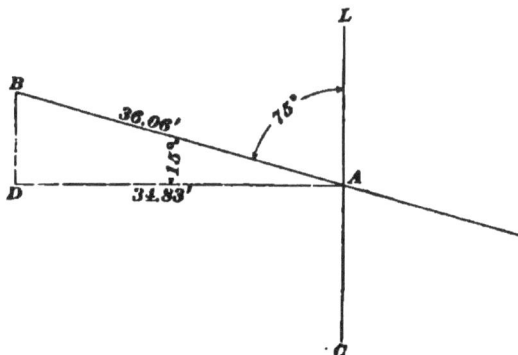

FIG. 76.

center line to the end of the culvert would be as follows: 21 ft. $-$ 5 ft. $=$ 16 ft.   16 ft. $\times 1\frac{1}{2} =$ 24 ft.   Adding 1.5 ft., and, in addition, 1 in. for each foot of embankment above the parapet, i. e., 16 in., or 1.33 ft., we have $24 + 1.5 + 1.33 = 26.83$ ft.   Adding 8 ft., one-half the width of the roadway, we have $26.83 + 8 = 34.83$ ft. $= A D$.   At D erect the perpendicular $D B$.   In the right-angled triangle $A D B$, we have $\cos A = \dfrac{A D}{A B}$, i. e., $\cos 15° = \dfrac{34.83}{A B}$; whence, $.96593 = \dfrac{34.83}{A B}$, and $A B = \dfrac{34.83}{.96593} = 36.06$ ft.   Ans.

(**898**)   See Art. **1541.**

(**899**)   See Art. **1541** and Fig. 462.

(**900**)   See Art. **1542.**

(**901**)   See Art. **1544.**

(902) The rod reading for grade will be the difference between the height of instrument and the elevation of the grade for the given station ; i. e., $125.5 - 118.7 = 6.8$ ft.

Ans.

(903) See Art. **1547.**

(904) See Art. **1548.**

(905) The skew of a bridge is the angle which its center line makes with the general direction of the channel spanned by the bridge.

(906) See Art. **1549.**

(907) The length of $A\,B$ is determined by the principles of trigonometry stated in Art. **1243,** from the following proportion : sin 46° 55′ : sin 43° 22′ :: 421.532 : $A\,B$; whence, $A\,B = 396.31$ ft.   Ans.

(908) See Art. **1551.**

(909) See Art. **1553.**

(910) See Art. **1554** and Figs. 465 and 466.

(911) The depth of the center of gravity of the water below the surface is $\frac{10}{2} = 5$ ft. Applying the law for *lateral pressure*, given in Art. **1554,** we have

(*a*) Total water pressure $= 10 \times 5 \times 40 \times 62.5 = 125,000$ lb.

Ans.

(*b*) Taking a section of the dam 1 ft. in length, we have

lateral pressure $= 10 \times 5 \times 62.5 = 3,125$ lb.

The *center* of water pressure is at $\frac{1}{3}$ the depth of the water above the bottom, i. e., at $10 \div 3 = 3\frac{1}{3}$ ft.

The moment of water pressure about the inner toe of the dam will, therefore, be $3,125 \times 3\frac{1}{3} = 10,417$ ft.-lb.   Ans.

(912) The volume of a section of the cofferdam 1 ft. in length is $11 \times 1 \times 5 = 55$ cu. ft., which, at 130 lb. per cubic foot, will weigh $55 \times 130 = 7,150$ lb.

(*a*) The moment of resistance of the cofferdam is the product of its weight, viz., 7,150 lb. multiplied by the perpendicular distance from the inner toe of the dam to the vertical line drawn through the center of gravity of the dam. This perpendicular distance is 2.5 ft. The moment of resistance is, therefore, $7,150 \times 2.5 = 17,875$ ft.-lb. Ans.

(*b*) The factor of safety of the cofferdam is the quotient of the moment of resistance of the dam divided by the moment of water pressure. Hence, $17,875 \div 10,417 = 1.71$. This calculation does not include the additional weight and resistance of the piles and timber enclosing the puddled wall, which will greatly increase the factor of safety.

(913) See Art. **1555.**

(914) See Art. **1555.**

(915) See Art. **1556.**

(916) See Art. **1557.**

(917) See Art. **1559.**

(918) See Arts. **1560** and **1561.**

(919) See Arts. **1563** and **1564.**

(920) As the sides of the bridge piers are always, and the ends often, battered, the deeper the foundation the greater will be the dimensions of the caisson plan. The thickness of the walls and deck of the caisson will depend upon the weight of the masonry and the bridge which they must support.

(921) To reduce the friction of the earth against its sides while sinking.

(922) See Art. **1566.**

(923) See Art. **1566.**

(924) See Art. **1568.**

(925) See Art. **1568.**

(926) See Art. **1570**

(927) Multiplying 70 ft., the depth of the water, by the decimal .434 and adding 15 lb., the pressure of the atmosphere, we have $70 \times .434 = 30.38$. $30.38 + 15 = 45.38$ lb.

Ans.

(928) Formula **109**, $L = \dfrac{2\,w\,h}{S+1}$. (See Art. **1572**.)

(929) See Art. **1573**.

(930) The striking force of the hammer is equal to 3,500 lb., the weight of the hammer multiplied by 30, the number of feet in its fall, or $3,500 \times 30 = 105,000$ ft.-lb.

Ans.

(931) See Art. **1575**.

(932) See Art. **1576**.

(933) See Art. **1577**.

(934) See Art. **1578**.

(935) See Art. **1580**.

(936) Applying formula **109**, $L = \dfrac{2\,w\,h}{S+1}$ (see Art. **1572**), in which $L =$ the safe load in tons, $w =$ the weight of the hammer in tons, $h =$ the height of the fall of the hammer in feet, and $S =$ the average penetration of the last three blows; we have, by substituting known values in the formula, safe load $L = \dfrac{3 \times 22}{1.5} = 44$ tons. Ans.

(937) Applying formula **109**, $L = \dfrac{2\,w\,h}{S+1}$, we find the safe load of each pile as follows: First pile, $L = \dfrac{3.3 \times 35}{.75 + 1} = 66$ tons; second pile, $L = \dfrac{3.3 \times 35}{.8 + 1} = 64.166$ tons; third pile, $L = \dfrac{3.3 \times 35}{.875 + 1} = 61.6$ tons; total safe load of three piles $= 191.77$ tons. Ans.

(938) See Art. **1585**.

(939) See Art. **1585.**

(940) See Art. **1586.**

(941) See Art. **1589** and Figs. 484, 485, and 486.

(942) See Art. **1590.**

(943) See Art. **1591**

(944) See Art. **1591.**

(945) See Art. **1591.**

(946) See Art. **1592.**

TRACK WORK.

(947) See Art. **1593.**

(948) See Art. **1595.**

(949) See Art. **1596.**

(950) See Art. **1597.**

(951) See Art. **1599.**

(952) See Art. **1600.**

(953) See Arts. **1602** and **1603.**

(954) See Art. **1606.**

(955) See Art. **1607.**

(956) See Art. **1608.**

(957) See Art. **1609.**

(958) The coefficient of expansion is .0000086 per degree per lineal foot (see Art. **1611**). As the temperature of the bar is 85° and normal temperature is 60°, the elongation per foot due to increase of temperature is equal to $85 - 60 = 25 \times .0000086 = .0001715$ of a foot for each foot of length of the bar, and the total elongation is equal to $.0001715 \times 30.016 = .00515$ ft., say .005 ft. $= \dfrac{1}{16}$ in. The normal length of the bar is, therefore, $30.016 - .005 = 30.011$ ft. Ans.

(959) See Art. **1611.**

(960) See Art. **1612.**

(961) See Art. **1615.**

(962) See Art. **1617.**

(963) See Art. **1618.**

(964) See Art. **1619.**

(965) See Art. **1619.**

(966) See Figs. 507, Art. **1618,** and 510, Art. **1619.**

(967) See Art. **1625.**

(968) See Art. **1631.**

(969) See Art. **1638.**

(970) See Art. **1642.**

(971) See Art. **1650** and Fig. 513

(972) See Art. **1650.**

(973) See Art. **1655.**

(974) See Art. **1656.**

(975) See Art. **1659.**

(976) See Art. **1662.**

(977) The distance between rail centers is 4 ft. 11 in., which, expressed in feet and the decimal of a foot, is 4.916 ft. The radius of an 8° curve is 716.78 ft. Applying rule **2,** under Art. **1664,** we have

$$\text{excess of length of outer rail} = \frac{4.916 \times 425}{716.78} = 2.91 \text{ ft.} \quad \text{Ans.}$$

(978) The chord $c = 30$ ft.; the radius $R = 955.37$ ft. Applying formula **112,** Art. **1665,**

$$m = \frac{c^2}{8\,R},$$

and substituting known values, we have

$$m = \frac{30^2}{8 \times 955.37} = \frac{900}{7,642.96} = .118 \text{ ft., nearly } 1\tfrac{1}{2} \text{ in.} \quad \text{Ans.}$$

(979) We find in Table 32, Art. **1665,** that the middle ordinate of a 50-ft. chord of a 1° curve is $\tfrac{5}{8}$ in. The middle ordinate of the given chord is $3\tfrac{1}{2}$ in., and the degree of the given curve is, therefore, the quotient, or, $3\frac{1}{4} \div \frac{3}{5} = 5.6° = 5° 36'$. The degree of the given curve is probably 5° 30'. Ans.

(980) See Art. **1667.**

(981) See Art. **1669.**

(982) Formula **113,** $c = 1.587\ V$. (See Art. **1670.**)

(983) Applying formula **113,** $c = 1.587\ V$ (see Art. **1670**), in which c is equal to the chord whose middle ordinate m is equal to the required elevation, and V the velocity of the train in miles per hour, we have

$$c = 1.587 \times 35 = 55.5\ \text{ft.}$$

We next find the middle ordinate m by applying formula **112,** Art. **1665,**

$$m = \frac{c^2}{8\,R},$$

in which $c = 55.5$ ft. and $R = 573.69$ ft. Substituting known values, we have

$$m = \frac{55.5^2}{8 \times 573.69} = \frac{3,080}{4,589} = .67\ \text{ft.} = 8\ \text{in., nearly.}\quad \text{Ans.}$$

(984) For each half inch of curve elevation we add one rail length to the elevated approach, which gives for a 4-in. elevation $4 \div \frac{1}{2} = 8$ rail lengths, equal to $8 \times 30 = 240$ ft. Ans.

(985) See Art. **1674.**

(986) See Art. **1676** and Fig. 522.

(987) See Art. **1679.**

(988) See Art. **1681** and Figs. 525, 526, and 527.

(989) See Art. **1682** and Fig. 528.

(990) See Art. **1683** and Fig. 529.

(991) See Art. **1684.**

(992) See Art. **1685** and Fig. 530.

(993) See Art. **1686** and Fig. 531.

(994)   See Art. **1687.**

(995)   See Art. 1687 and Fig. 532.

(996)   See Art. **1688** and Fig. 533.

(997)   See Art. **1689** and Fig. 535.

(998)   See Art. **1690** and Fig. 536.

(999)   See Art. **1691** and Fig. 538.

(1000)   Applying formula **115,** given in Art. **1692,**

frog number $= \sqrt{\text{radius} \div \text{twice the gauge}}$,

and substituting known values, we have

frog number $= \sqrt{602.8 \div 9.417}$; whence,
frog number $= 8$, almost exactly.   Ans.

(1001)   Applying formula **117,** given in Art. **1692,**

radius $=$ twice the gauge $\times$ square of frog number,

and substituting known values, we have

radius $= (4 \text{ ft.} 8\frac{1}{4} \text{ in.}) \times 2 \times 6^2 = 9.417 \times 36 = 339 \text{ ft.}$
Ans.

To find the degree of the required curve, we divide 5,730 ft., the radius of a 1° curve, by 339, the length of the required radius, which gives $16.902° = 16° 54'$, the degree of the required curve.

(1002)   From table of Tangent and Chord Deflections, we find the tangent deflection of an 18° curve is 15.64 ft. Calling the required frog distance x, we have, from Art. **1692** and Fig. 540, the following proportion:

$$15.64 : 4.75 :: 100^2 : x^2;$$

whence, $x^2 = \dfrac{100^2 \times 4.75}{15.64} = \dfrac{47,500}{15.64} = 3{,}037.08,$

and $x = 55.1 \text{ ft.}$ Ans.

The frog angle is equal to the central angle of a 55.1 ft. arc of an 18° curve.   As 18° is the central angle of a 100-ft. arc, the central angle of an arc of 55.1 ft. is $\dfrac{55.1}{100}$ of 18° $=$ 9° 55'.   Ans.

(1003) This example comes under Case I of Art. **1693.** The sum of the tangent deflections of the two curves for chords of 100 ft. is equal to the tangent deflection of a 16° 30′ curve for a chord of 100 ft. This deflection, we find from the table of Tangent and Chord Deflections, is 14.35 ft. Calling the required frog distance x, we have, from Art. **1692,** the following proportion:

$$14.35 : 4.75 :: 100^2 : x^2;$$

whence, $\quad x^2 = \dfrac{100^2 \times 4.75}{14.35} = \dfrac{47,500}{14.35} = 3,310,$

and $\quad x = 57.5$ ft. Ans.

As the central angle subtended by a 100-ft. chord is 16° 30′, the angle subtended by a chord of 1 ft. is $\dfrac{16° \; 30′}{100} = 9.9′$, and for a chord of 57.5 ft. the central angle is $9.9′ \times 57.5 = 569.25′ = 9° \; 29\frac{1}{4}′$, the required frog angle.

(1004) This example comes under Case II of Art. **1693,** in which the curves deflect in the same direction. As the main track curve is 4° and the turnout curve 12°, the rate of their deflection from each other is equal to the deflection of a 12° − 4°, or of an 8° curve from a straight line. From the table of Tangent and Chord Deflections, we find the tangent deflection of an 8° curve for a chord of 100 ft. is 6.98 ft. Now, the distance between the gauge lines at the head-block is fixed at 5 in. = .42 ft., and calling the required head-block distance x, we have the proportion

$$6.98 : 0.42 :: 100^2 : x^2;$$

whence, $\quad x^2 = 601.7$, and

head-block distance $x = 24.5$ ft. Ans.

(1005) Multiply the spread of the heel by the number of the frog; the product will be the distance from the heel to the theoretical point of frog.

(1006) See Art. **1694.**

(1007) $60 \times 12 = 720$ in. $720 \div 15 = 48.$ Ans.

(1008) 8 ft. 6 in. = 102 in.   15 ft. 3 in. = 183 in.   183 in. − 102 in. = 81 in.   81 in. ÷ 48 = 1.69 in. = $1\frac{11}{16}$ in. Ans.

(1009) See Art. **1698.**

(1010) We find the crotch frog distance as follows: From the table of Tangent and Chord Deflections, we find the tangent deflection of a 100-ft. chord of a 10° curve is 8.72 ft. One-half the gauge is 2.35 ft. Calling the crotch frog distance x, we have the proportion

$$8.72 : 2.35 :: 100^2 : x^2;$$

whence, $x^2 = \dfrac{100^2 \times 2.35}{8.72} = 2,695$, nearly.

and $x = 51.9$ ft.   Ans.

The central angle for a chord of 51.9 ft. is $6' \times 51.9 = 311.4' = 5° 11.4'$, and the angle of the crotch frog is $5° 11.4' \times 2 = 10° 22.8'$. Ans.

(1011) See Art. **1700.**

(1012) See Art. **1710.**

(1013) See Art. **1711.**

(1014) See Art. **1712.**

(1015) See Art. **1714.**

(1016) See Art. **1715.**

RAILROAD STRUCTURES.

(1017) Eight years.

(1018) See Art. **1731.**

(1019) See Art. **1732.**

(1020) See Art. **1734.**

(1021) See Art. **1734.**

(1022) See Art. **1735.**

(1023) Test piles should be driven at frequent intervals and the required lengths determined by actual measurement.

(1024) See Art. **1739** and Fig. 558.

(1025) See Art. **1739.**

(1026) See Art. **1741** and Fig. 561.

(1027) See Art. **1742.**

(1028) See Art. **1742** and Figs. 562 to 567, inclusive.

(1029) See Art. **1745** and Fig. 565.

(1030) See Art. **1746** and Fig. 566.

(1031) See Art. **1748** and Fig. 567.

(1032) See Art. **1749** and Figs. 568 and 569.

(1033) See Art. **1752** and Figs. 577 and 578.

(1034) See Art. **1753** and Figs. 579 to 584, inclusive.

(1035) See Art. **1754** and Figs. 585 to 587, inclusive.

(1036) See Art. **1756** and Figs. 588 to 591, inclusive.

(1037) See Art. **1756** and Fig. 592.

(1038) See Art. **1757.**

(1039) See Art. **1758** and Fig. 592.

(1040) See Art. **1759.**

(1041) See Art. **1759.**

(1042) See Art. **1760.**

(1043) See Art. **1760** and Figs. 598 to 601, inclusive.

(1044) See Art. **1761** and Fig. 604.

(1045) See Art. **1763** and Figs. 606 and 607.

(1046) See Art. **1767** and Figs. 608, 609, and 610.

(1047) See Art. **1770.**

(1048) See Art. **1775** and Figs. 617 and 618.

(1049) See Art. **1778.**

(1050) See Art. **1780.**

(1051) See Art. **1787.**

(1052) See Art. **1789** and Fig. 622.

(1053) See Art. **1800.** From Table 52, Art. **1802,** we find the constant for cross-breaking center loads for spruce is 450. Applying formula **127,** Art. **1800,** we have

$$\frac{10 \times 144}{18} \times 450 = 36,000 \text{ lb.} \quad \text{Ans.}$$

(1054) See Art. **1803.**

(1055) The constant for center breaking loads for yellow pine is 550, Table 52, Art. **1802.** Applying the rule given in Art. **1804,** we have $\frac{50,000 \times 20}{550} = 1,818.$ The cube root of 1,818 is 12.2 nearly. Hence, the beam should be 12¼ inches square. Ans.

On account of the great size of this beam, its own weight constitutes a considerable factor of the breaking load, and to provide for this additional weight we must increase the size of the beam as directed in Art. **1804.**

Yellow pine weighs, on an average, 65 lb. per cu. ft. The beam between supports contains 20.8 cu. ft. Its weight is, therefore, $20.8 \times 65 = 1,352$ lb. Applying the proportion given in Art. **1804,** and denoting the required addition in width by x, we have

$$50,000 : \frac{1,352}{2} :: 12.2 : x;$$

whence, $x = 0.16$ in. The beam should, therefore, be $12.2 + 0.16 = 12.36$ inches square, equal to about $12\frac{3}{8}$ inches.

(**1056**) A safe center load of 16,000 lb. with a factor of safety of 5, will call for a center breaking load of $16,000 \times 5 = 80,000$ lb. We next find, by Art. **1804,** the side of a square beam which will break under this center load of 80,000 lb., and we have $\dfrac{80,000 \times 16}{550} = 2,327$, the cube root of which is 13.25, nearly. Hence, the side of the required square beam is $13\frac{1}{4}$ inches. Ans.

(**1057**) The constant for center breaking loads for spruce is 450. (See Table 52, Art. **1802.**) Applying the rule given in Art. **1806,** we have

$$\frac{40,000 \times 14}{144 \times 450} = \frac{560,000}{64,800} = 8.64 \text{ in.} = 8\frac{5}{8} \text{ in.} \quad \text{Ans.}$$

(**1058**) The constant for center breaking loads for yellow pine (Table 52, Art. **1802**) is 550. Applying the rule given in Art. **1807,** we have

$$\frac{24,000 \times 16}{10 \times 550} = \frac{384,000}{5,500} = 69.82,$$

the square root of which is $8.35 = \text{say } 8\frac{3}{8}$, the required depth in inches of the beam.

(**1059**) Applying formula **128,** Art. **1808,** we nave:

Breaking load in pounds per square inch of area =

$$\frac{5,000}{1 + \left(\dfrac{144^2}{12^2} \times .004\right)} = 3,172 \text{ lb.}$$

With a factor of safety of 5, we have for the safe load per square inch, $\dfrac{3,172}{5} = 634$ lb. The area of the cross-section of the pillar is $12 \times 12 = 144$. Hence, the safe load for the pillar is $144 \times 634 = 91,296$ lb. Ans.

(**1060**)   From Art. **1809** we find the safe shearing stress across the grain for long-leaf yellow pine to be 500 lb. per sq. in., and to resist a shearing stress of 30,000 lb. will require an area of $\frac{30,000}{500} = 60$ sq. in.   Ans.

(**1061**)   From Art. **1809** we find the safe shearing stress across the grain for white oak to be 1,000 lb. per sq. in., and to sustain a shearing stress of 40,000 lb. will require an area of $\frac{40,000}{1,000} = 40$ sq. in.   Ans.

(**1062**)   A uniform transverse safe load of 36,000 lb. is equivalent to a center load of 18,000 lb., which, with a factor of safety of 4, is equivalent to a center breaking load of 72,000 lb. $72,000 \times 12 = 864,000$.   The constant for transverse breaking loads for spruce is 450. (Table 52, Art. **1802**.) $\frac{864,000}{450} = 1,920$, the cube root of which is 12.4, the side dimension in inches of a square beam, which will safely bear the given transverse load.   (Art. **1805**.)

For tension, we find in Art. **1810** the safe working stress for ordinary bridge timber is 3,000 lb. per sq. in. The given beam has a pulling stress of 20,000 lb., and to resist this stress it will require $\frac{20,000}{3,000} = 6.6$ sq. in.   We must, accordingly, add this to the area of the beam required to sustain the transverse load alone.   We determine the increased size as follows: $\sqrt{12.4^2 + 6.6} = 12.6$;   hence, the beam is 12.6 inches square.   Ans.

(**1063**)   See Art. **1813** and Fig. 629.

(**1064**)   The total load upon the bridge is $6,000 \times 20 = 120,000$ lb.   Of this amount the two king-rods sustain one-half, or 60,000 lb., which places upon each king-rod a load of $\frac{60,000}{2} = 30,000$.   This, with a factor of safety of 6, would be equivalent to an ultimate or breaking load of $30,000 \times 6 = 180,000$ lb.   By reference to Table 53, Art. **1813**, we find that a rod 2⅜ in. in diameter has a breaking strain of 178,528 lb., which is a close approximation to the given stress. We would, accordingly, use a 2⅜ in. king-rod. Ans.

(1065) The king-rods support the needle-beam, and, as all the floor-beams rest upon the needle-beam, it carries half the total bridge load excepting what rests upon the tie-beams.

(1066) By trussing the needle-beam we practically reduce its span to one-half of its actual length.

(1067) See Art. **1814** and Fig. 631.

(1068) 20,000 lb. Ans.

(1069) The load per lineal foot on each truss is $\frac{6,000}{2} =$ 3,000 lb. As each rod supports half the load between itself and each adjacent support, it will support one-third of the total load of each truss. The total load of the truss is 3,000 × 33 = 99,000 lb., one-third of which is $\frac{99,000}{3} =$ 33,000 lb. (See Art. **1814**.) Ans.

(1070) See Art. **1815.**

(1071) See Art. **1815.**

(1072) See Art. **1816.**

(1073) See Art. **1816.**

(1074) See Art. **1816.**

(1075) See Art. **1817.**

(1076) See Art. **1818.**

(1077) See Art. **1818.**

(1078) See Art. **1819** and Fig. 636.

(1079) See Art. **1820.**

(1080) See Art. **1821.**

(1081) See Art. **1822.**

(1082) See Art. **1823.**

.